Skeptical Environmentalism

Skeptical
Environmentalism

The Limits of Philosophy and Science

Robert Kirkman

Indiana University Press

BLOOMINGTON & INDIANAPOLIS

This book is a publication of

Indiana University Press
601 North Morton Street
Bloomington, IN 47404-3797 USA

http://iupress.indiana.edu

Telephone orders   800-842-6796
Fax orders   812-855-7931
Orders by e-mail   iuporder@indiana.edu

The paper used in this publication meets the minimum requirements of
American National Standard for Information Sciences—Permanence of Paper
for Printed Library Materials, ANSI Z39.48-1984.

Manufactured in the United States of America

Library of Congress Cataloging-in-Publication Data

Kirkman, Robert, date
    Skeptical environmentalism : the limits of philosophy and science /
Robert Kirkman.
        p.   cm.
Includes bibliographical references and index.
    ISBN 0-253-34037-3 (alk. paper) — ISBN 0-253-21497-1 (pbk. : alk.
paper)
    1. Environmentalism—Philosophy. 2. Environmental ethics.
3. Environmental responsibility. I. Title.
    GE195 .K57 2002
    304.2'01—dc21

                                                    2001002951

1 2 3 4 5   07 06 05 04 03 02

For Andrea

Contents

Acknowledgments

No book ever grows in a vacuum. The seeds for *Skeptical Environmentalism* were planted in 1991, when I was a first-year doctoral student in philosophy at the State University of New York at Stony Brook; their first growth became my doctoral dissertation, which I defended in November 1995. I owe more thanks than I can express to my advisors and teachers, especially Ed Casey, Mary Rawlinson, Lawrence Slobodkin, Marsh Spector, Walter Watson, and Anthony Weston.

From 1990–1992, I had the privilege of participating in an environmental philosophy reading group with my fellow graduate students Cas Ivanbrook, Irene Klaver, Dave Macauley, and Chuck Wright; a counterpoint was provided by our graduate student colleagues in the Department of Ecology and Evolution at Stony Brook: Sean Craig, Paul Fernhout, Siana LaForest, and Tim Morton. They also have my gratitude; our many discussions shaped my thinking in ways that continue to surprise me.

Much of the research for Chapters 2 and 3 was conducted in France during the 1993–1994 academic year with the support of a more-than-generous Chateaubriand Scholarship from the Cultural Consul of the French Embassy to the United States. Anthony Steinbock, François Raffoul, François Dagognet, and Georges Canguilhem deserve special thanks for their help in getting me to France, as do Pascal Acot and Jean-Marc Drouin for their help and guidance during my stay. Some of the material in Chapters 2 and 3 was published, in a slightly different form, in the journal *Envi-*

ronmental Ethics; I am grateful to Gene Hargrove and two anonymous readers for their comments.

In addition, a small portion of Chapter 4 was previously published as part of a book review in *The Journal of Value Inquiry*; I am grateful to Kluwer Academic Publishers for permission to reuse that material.

Thanks also go to Andrew Brennan, Drew Christie, Robert McIntosh, Bob Scharff, Jim Snyder, and Steve Vogel for their help along the way; to Bruce Foltz and another anonymous reader for invaluable encouragement and criticism on the near-final manuscript.

I am, as always, grateful to my wife for her patience, sympathy, companionship, and no-nonsense editorial advice. This book is dedicated to her.

Skeptical Environmentalism

Introduction

Skepticism is a dangerous business, not least because it is so easily misunderstood. In the final section of his *Enquiry Concerning Human Understanding*, David Hume noted that the skeptic, along with the "speculative atheist," has traditionally been identified as an enemy of religion and so "naturally provokes the indignation of divines and graver philosophers." More broadly, skepticism is often taken to be a flat denial that any knowledge is possible, which raises the specter of the most abject rootless relativism. Hume went on to say, however, that the kind of skeptic that haunts the dreams of divines and philosophers does not exist. "No man ever met with any such absurd creature," he writes, "or conversed with a man, who had no opinion or principle concerning any subject or speculation." For Hume, this left two questions: "What is meant by a sceptic? And how far is it possible to push these philosophical principles of doubt and uncertainty?"[1]

The first thing to know about this book, then, is that I follow Hume in his rejection of extreme forms of skepticism in favor of a moderate or "mitigated" skepticism. I even go so far as to adopt Hume's two principles of moderate skepticism, with some qualifications. The human mind tends toward dogmatism, Hume noted, but when we are shown the "infirmities" of human understanding, we are naturally inspired to "more modesty and reserve." This led him to his first principle: "in general, there is a degree of doubt, and caution, and modesty, which, in all kinds of scrutiny, ought for ever to accompany a just reasoner." Further, the human imagination tends to soar to the very boundaries of the universe, against which tendency Hume offered his second principle: a correct judgment "avoids all distant and high enquiries, conforming itself to common life, and to such subjects as fall under daily practice and experience." Thus, he advised, we should limit "our enquiries to such objects as are best adapted to the narrow capacity of human understanding."[2]

While I agree with these principles in broad outline, I do not agree with the standards by which Hume himself assessed claims to knowledge. As an empiricist, Hume believed that all knowledge begins with experience, which he understood as "impressions" left on the mind by the senses, much as marks are left on a chalkboard by the impact of chalk. The mind itself is passive and contributes nothing new to knowledge. On this basis he argued, for example, that humans can have no necessary knowledge of cause and effect. No matter how many times he saw one event following another, Hume insisted, he never saw the causal connection between them. Because his model of consciousness excludes any other source of knowledge, Hume was left with nothing more than the constant conjunction of the events accompanied by a habitual expectation that some combinations of events somehow belong together. This conclusion makes sense only within a particularly narrow conception of human consciousness, and a more thoroughgoing skepticism would call even this conception into doubt.

In addition to this qualification, I would introduce a third principle to govern how these first two principles are applied: the principle of parity. In the public arena, skepticism can be a powerful weapon to use against intellectual and political adversaries, particularly when deeply held values are at stake. In the highly competitive interest-group politics of our times, environmentalists are only too happy to raise doubts about the status quo or about the interests and arguments of their opponents. For their part, opponents of environmentalism enthusiastically return the favor. Often enough, they take up the banner of skepticism as they lay siege to this or that environmentalist claim. To win the hearts and minds of the public, advocates for all sides maintain that they alone have the best science, the purest motives, and the soundest policies and that their opponents wallow in irrationality, corruption, and folly.

The principle of parity holds that skeptical principles should be applied more even-handedly. As Hume might put it, the human mind has a strong tendency to remain convinced of its own rightness—or even righteousness—in spite of evidence to the contrary. Especially when the stakes are high, humans are far less likely to scrutinize their own beliefs than those with which they disagree. When scrutiny is brought to bear and deeply held beliefs are found wanting, the mind has an almost unlimited capacity for evasion and denial. The principle of parity is designed as a corrective against denial and the very real dangers it can entail. In the context of this book, the principle can be stated as follows: let those of us who identify ourselves as environmentalists examine our own assumptions and practices with at least the same degree of critical candor we bring to bear on the assumptions and practices of the status quo.

Informed by these three principles, the basic idea of a skeptical environmentalism is simply stated: *all* thought and discussion about human life in its natural environment ought to be accompanied by a measure of doubt and modesty and a concern to stay within the limits of what is possible for human knowledge.

Speculative Environmentalism

This present work is focused fairly narrowly. Although there is a crying need for fair and consistent application of skepticism in the public arena, I am not concerned so much with the debate over concrete matters of policy as I am with the more basic principles that inform the work of many environmentalists. Ethical principles are involved, certainly, but these are often placed in a particular metaphysical or cosmological context. Many environmentalists appeal, tacitly or explicitly, to their belief in a fundamental connectedness or relatedness in the natural world that extends to and encompasses Homo sapiens, an ecological worldview that supports and justifies environmentalist policies.

Still more narrowly, I am interested in the work of academic environmental philosophers who, since the early 1970s, have taken it upon themselves to formalize the core beliefs of the environmental movement. Many of them are apparently motivated by the hope that if the environmental movement can be united behind a common set of principles and if those principles are grounded in the best that philosophy and the natural sciences have to offer, then the movement is much more likely to prevail in the political arena. My first task will be to determine whether environmental philosophers are likely to succeed in establishing such a groundwork, given the limits of knowledge.

This task is complicated by the fact that environmental philosophy is diverse: it has not yielded just one set of environmentalist principles, but many. The field encompasses those who call themselves environmental ethicists, deep ecologists, ecofeminists, social ecologists, postmodern environmentalists, environmental pragmatists, and others, all of whom seem to be pulling in different directions. I believe that, to some extent at least, this apparent diversity masks a deeper unity. Environmental philosophers have tended to pursue the same basic projects in more or less the same manner, even as they disagree over details. In this sense, they have produced a set of variations on a common theme.

I could simplify matters by focusing entirely on the theme itself in the abstract. In doing so, however, I would risk doing violence both to the work of individual environmental philosophers and to the credibility of my own argument. Instead, I will attempt to strike a balance between the need to generalize and the need to acknowledge particulars. So while I will reconstruct what I take to be the common theme of environmental philosophy, I will also consider a range of variations wherever it seems appropriate to do so. I will also note the extent to which some forms of environmental philosophy—including recent work informed by postmodernism, feminism, and pragmatism—have bucked the trend altogether, setting off in a very different direction.

At the heart of much environmental philosophy I distinguish three general projects, which I call the practical, the radical, and the speculative projects respectively. The practical project is the most basic and the most widely shared. Environmental philosophers are generally concerned with finding some way of making sound decisions in the face of the perceived environmental crisis. Many also seem to be convinced that the best practical solutions are to be found by intellectual means.

It is a commonplace among environmental thinkers that the root cause of environmental problems is a way of thinking that alienates humans from their natural environment and so permits or even encourages an accelerating disruption of natural systems. What I would call the radical project of environmental thought is an effort to identify, expose, and root out this destructive way of thinking. In its strongest form, this project expresses itself in what Anna Bramwell called "Manichean" ecology, after the early variant of Christianity that held that good and evil were equal powers in the universe engaged in a never-ending struggle for supremacy. In her history of "ecology," understood as a political ideology, Bramwell described Manichean ecologists as holding to the view that humans are natural beings and thus ought to occupy a natural niche, but that humans are now, in fact, behaving unnaturally and are living in alienation from nature.

Given the paradox that natural man behaves unnaturally, what went wrong? Various explanations put forward have in common the tendency to point to a guilty party. There are several different guilty parties in common usage. These are Christianity, the Enlightenment (with atheism, scepticism, rationalism and scientism following on), the scientific revolution (incorporating capitalism and utilitarianism), Judaism (via either the Jewish element in Christianity or via capitalism), Men, the Nazis, the West, and various wrong spirits, such as greed, materialism, acquisitiveness, and not knowing where to stop.

She went on to note that it is very difficult to pin down the roots of anti-nature sentiment, as any such analysis "suffers from a bewildering variety of historical locations and cruces."[3] I would add that academic environmental philosophy has contributed liberally to this bewildering variety, including David Abram's recent argument that alphabetic writing is at least one of the reasons that "nonhuman nature seems to have withdrawn from both our speaking and our senses."[4]

Whatever the importance of the radical project, many environmental philosophers take it as their primary task to replace the destructive way of thinking with one that is more benign, an ecological worldview that embodies a conception of *relatedness*. The search for such a view constitutes what I call the speculative project of environmental philosophy—"speculative environmentalism" for short. An appropriate ecological worldview usually encompasses three broad claims about human life in its environmental context, which, together, make up the central case of speculative environmentalism:[5]

- The natural world is fundamentally relational.
- Humans have a moral obligation to respect and preserve the (relational) order of nature.
- Widespread acceptance of the first two claims is the key to solving the environmental crisis.

The first claim is offered in direct contradiction to the mechanistic view of nature attributed to Descartes and Newton, which is seen to reduce the non-human world to a mere collection of isolated physical entities with no value or purpose of their own. This mechanistic view is often identified as the root cause of environmental ills because it seems to license the wholesale rearrangement of the collection to suit human purposes. By contrast, an ecological worldview begins with the belief that the natural order is more like an organism, a web of interconnections that has its own internal unity, its own goals and interests. As such, the natural order can be harmed. All of this presupposes the possibility of some degree of knowledge of nature and its interests, whether the source of this knowledge is speculation or science. In Part 1, I address this presupposition, arguing that neither speculative reason nor scientific inquiry lends unambiguous support to an ecological worldview.

The second claim, regarding human obligations, is generally taken as following from the first claim in one of two ways. If nature is an organism with interests of its own, then natural entities and systems may have a significance or value far beyond their usefulness for humans. Further, if humans participate in the system of relations, then they may have obligations to the system just as citizens have obligations to the nation of which they are a part. I argue in Part 2 that even if a relational worldview can be established beyond doubt, moral obligations do not follow simply as a matter of course.

The third claim expresses the basic hope of environmental philosophy. If the roots of the environmental crisis are intellectual, then the solution must also be intellectual. Philosophers would thus be raised to a privileged position in the debate over human-induced environmental change, since they would have the power to undo the intellectual or spiritual damage that was done by whatever it is that alienated humans from nature in the first place. I call this hope into question in Part 3, where I consider the possibility that there may be no solution to the environmental crisis at all, let alone one as tidy as a system of speculative philosophy.

If my reconstruction of the core projects of environmental philosophy is correct, then disagreements among environmental philosophers arise for two reasons: either they disagree about the relative importance of the three projects or they disagree about the manner in which one or another of the projects is to be carried out. Environmental ethicists differ from deep ecologists, for example, not only in their general approach to the speculative project, but in that deep ecologists tend to emphasize the radical project more strongly than do environmental ethicists. Among environmental ethicists, there is also disagreement over which metaphysic of morals, if any, is needed to support a truly environmental ethic. This is often accompanied by a disagreement over the implications of ecology. Some postmodernists, ecofeminists, and social ecologists claim to reject the speculative project out of hand, focusing almost exclusively on the radical and practical projects—even though it is not at all clear that the radical and speculative projects are so easily separated.

THE APPEAL TO HISTORY

I have set out to raise doubts about environmental philosophy, and much of what follows is a direct critical examination of the work of environmental philosophers. However, I have also drawn extensively from the history of philosophy and from the history and philosophy of science, including sources that do not seem to be directly relevant to environmental concerns. Chapter 2, for example, includes a lengthy discussion of the development of speculative philosophy since Descartes, which is crowned by a discussion of Hegel's *Philosophy of Nature.* This has struck a number of my colleagues as odd, and I should give some account of why I have done this.

The second principle of skepticism holds that anyone involved in any sort of inquiry should be careful to stay within the limits of human understanding. But where are those limits? One way to get at an answer to this question is to consider the work of past philosophers as they draw out the consequences of their basic principles in order to see whether, where, and why their efforts begin to break down. It is reasonable to suppose that if more recent thinkers were to start with similar principles and head in a similar direction, they might have similar results. For example, if the impulses of romantic nature philosophy ran into difficulties in the nineteenth century, is it likely they will they do better in the twenty-first? I do not expect that a historical study of this kind will produce a conclusive argument against the speculative tendencies of environmental philosophy; the appeal to history does, however, bring to light some important reasons for doubt.

There is a further justification for the historical bent of my argument, which is that the relationship of environmental philosophy to the philosophical and scientific tradition is itself in question. Hegel becomes relevant in Chapter 2 because some environmental philosophers have made an explicit appeal to romantic nature philosophy to inform or inspire their work. Many more appeal to the work of scientists, especially ecologists, for this same purpose. To determine whether these appeals accomplish what they are meant to accomplish (i.e., adding to the intellectual authority of an ecological worldview), it is necessary to look at the sources in some detail. For instance, I argue in Chapter 4 that many environmental thinkers misappropriate elements of scientific theory, but this can only come to light through a detailed discussion of the development of ecology as a distinct scientific discipline.

WHY I WROTE THIS BOOK

Before I get on with the task at hand, I want to be sure that my motives are well understood. To that end, let me state clearly what skeptical environmentalism is not. First, it is not an attack on environmentalism as such, nor is it an attempt to undermine and silence critics of the status quo. I am raising questions about one aspect of the environmental movement; namely, the intellectual habits of many academic environmental philosophers. Surely the movement as a whole is diverse and robust enough to survive such scrutiny—and perhaps even be the better for it. Further, the principle of parity demands that skeptical principles be applied consistently: it is both possible and advisable to subject the beliefs and practices of both the status quo and its critics to careful and sustained scrutiny.

Second, and more narrowly, skeptical environmentalism is not an attack on environmental philosophy as such. Although I mistrust labels, I usually refer to myself as an environmental philosopher in the broadest sense: I am a philosopher who brings the tools of my trade to bear on the complex environmental problems that confront humanity. I am convinced that philosophers have a great deal to contribute to the clarification and resolution of these problems. And yet I have long been uneasy with the work of many environmental philosophers; or, to be more precise, I have been uneasy with the way in which many environmental philosophers have gone about their work. It took me a while to pinpoint the source of my uneasiness. After several false starts, it occurred to me that environmental philosophers pursue not one but several different projects—introduced above as the practical, radical, and speculative projects—each of which could be assessed on its own merits. From this I concluded that it should be possible to criticize, limit, or even reject outright one of the projects—the speculative project in particular—without rejecting environmental philosophy as such or negating many of its hard-won insights.

Third, skeptical environmentalism is not entirely negative. Granted, much of what follows consists of argument after argument against speculative environmentalism, leading to the conclusion that environmental philosophers may be well advised to abandon the speculative project altogether. This conclusion raises a further question: If environmental philosophers are to abandon the speculative project, what should they be doing instead? I take this question seriously. The point of skeptical environmentalism is not to cripple environmental philosophy by tearing away any part of it that seems less than perfect; the point is to transform environmental philosophy by opening up new possibilities. In Chapter 5, I will begin the shift to a more positive or constructive phase of skeptical environmentalism, indicating some promising directions for environmental thought within the bounds of caution and modesty. In all fairness, I must once more recognize that a number of other environmental philosophers have already left the speculative project behind and struck out on their own. I am making just one contribution to a process that is already underway.

Beneath these public motives for writing this book lies a more personal motive. I am an environmentalist. Humans, especially those of us who live in "advanced" nations, have been transforming our environment to suit what we take to be our own best interests. We cannot always predict the consequences of the changes we make, especially in the long term, and we do not always agree about where our best interests lie. Too often, I believe, the mad rush to generate material wealth has only impoverished human life in the world and, increasingly, placed human life itself in danger. I also believe that the majority of people in the society of which I am a part—including those in positions of power—are either in ignorance of or in denial about the consequences of environmental change. As an environmentalist, I do what I can to foster ways of thinking and valuing and acting that will enrich human life and help to ensure its sustainability.

I am also a skeptic. I do not easily give my allegiance to any one doctrine or principle. I mistrust labels and easy categorization. I continually question motives and assumptions, especially my own. In reconstructing the history of my intellectual development, I sometimes appeal to the Jeffersonian image of periodic revolution: since adolescence, I have made a point of allowing my entire way of thinking to be overturned from time to time, simply to avoid complacency. As a consequence, I am far from a model activist.

This book is a product of my ongoing struggle to reconcile these two facets of my character, to discover how and to what extent it is possible to be both a skeptic and an environmentalist at the same time.

Part 1

Knowledge

[1]

The Nature of Nature

Environmentalism will succeed only if its advocates can bring about a change in the way people behave. How can environmentalists do this? Answers come from all sides: regulate, legislate, litigate, negotiate, innovate, and educate; restructure the marketplace to create new incentives; restructure the schools to create a new kind of citizen; restructure civilization itself. In the midst of all these possibilities, environmental philosophy began with the belief that the best way to change the way people behave is to change the way they think. Not just any change would do. By and large, environmental philosophers have not been content to tinker with momentary opinions on matters of politics and economics. Instead, they have insisted that people rethink their answers to the most fundamental questions of human life in the world: What is the nature of nature, and what is my place within it?

What is of value, and what are my obligations? For what may I hope?

So the search is on for a way of thinking about nature, about the cosmos, about reality itself that might fundamentally alter the ethical and political life of modern civilization. Many of those engaged in the search think of themselves as constituting a minority tradition, swimming against the intellectual current of modernity: in opposition to the fractured metaphysics of René Descartes, which they see as having set humans at odds with nature and with themselves, they propose an ecological worldview informed by a vision of relatedness. If nature is fundamentally relational, and if humans are caught up in those relations, then there may be some metaphysical leverage for ethical obligations toward nature. If humans have obligations toward nature, then there may be some ethical leverage for better public policies regarding environmental change. This, at least, is the hope of speculative environmentalists.

Descartes is important because he established the major problems of modern philosophy, contributed to the development of the modern sciences, and generally set the tone for the modern era. Critics charge that his doctrine of dualism introduced a schism into human thought, a wound that has not yet healed. Cartesian dualism holds that the world consists of two kinds of substance: mental substance, which thinks but takes up no space, and material substance, which takes up space and does not think. If Descartes is correct that material substance does nothing but take up space, it follows that material bodies can relate to each other only spatially. Suppose a number of billiard balls are resting on a pool table and that the cue ball is two feet away from the eight ball (proximity). I roll the cue ball toward the eight ball (relative motion), the cue ball strikes the eight ball (direct contact), and the two balls move off in different directions (relative motion again). Once these spatial relations—proximity, relative motion, and contact—have been measured and catalogued, there is nothing more to be learned about the situation on the pool table. In the Cartesian universe, the same applies to stars and planets, rocks and rivers, plants and animals, and even the human body.

For complex material objects such as animals, though, another feature of the dualistic conception of material substance comes into play. Matter is divisible, and so reduction becomes the proper method for studying complex systems. If I want to understand a clock, I need to disassemble it, study all of the parts, and account for their relationships to one another—in spatial terms, of course. The same holds if I want to understand a maple tree, a domestic cat, the workings of the human brain, or the cosmos as a whole. In effect, the Cartesian cosmos can be thought of as a great machine, designed and set in motion by a very powerful—and very clever—mechanic. Parts may come into contact with one another, and they may move relative to one another, but each can be understood in isolation from the others and each can be replaced if necessary.

A number of metaphysical and epistemological problems are associated with dualism, not least the problem of where to put the mind. If the mind takes up no space, how can it be located *in* space, in relation to a material body? This is an interesting puzzle, but many critics are more concerned with what they see as the pernicious ethical and political consequences of dualism—especially from an environmentalist point of view. When people see the natural world as a collection of material bodies, these critics charge, they come to treat the collection as nothing more than a stockpile of resources for human consumption. At the same time, reductionism and mechanism have allowed modern science to discover some of the inner workings of phenomena, adding to the variety of spatial and material relations to be catalogued and making possible more and more extensive and intrusive alterations of natural systems. Over all of this is spread an abiding modernist faith in the ability of humanity to solve its problems by rational, especially technological, means, which ensures the unending progress of human civilization as it conquers brute nature and secures its own future. Critics fear that this dominant paradigm (as some call it) of the modern era is a recipe for environmental disaster: civilization advances under Descartes' banner, blind to the destruction it leaves in its wake.

Those who call themselves deep ecologists are among the most strident critics of the dominant paradigm. Deep ecology is a political and philosophical movement first introduced into English-language philosophy by Norwegian philosopher Arne Naess in 1973. Naess distinguishes two different kinds of "ecology," two branches of the environmental movement: the shallow and the deep. Shallow ecology remains within the limits of the dominant paradigm; it simply adds some recognition that resources are not inexhaustible and that the continuing progress of civilization may require a good deal of prudence. The goal of shallow ecology, as Naess portrays it, is to fight pollution and resource depletion in order to ensure "the health and affluence of people in developed countries." It is "shallow" because it seeks to reform the system without challenging the dominant paradigm or the social and economic system to which it has given rise. Naess singles out the "man-in-environment image" as central to the worldview of the shallow ecologist: humans live out their lives against a neutral backdrop of material resources. He rejected this image in favor of a "relational, total-field image."[1] The foremost American interpreters of Naess's ideas, Bill Devall and George Sessions, argue that the task of deep ecology is to reverse what sociologist Max Weber called the "disenchantment of the world" which was brought about by the rise of "instrumental rationality."[2]

The way to reverse the destructive tendencies of modern civilization, according to many environmental philosophers, is to view nature instead as a kind of organism. Unlike machines, organic systems are so tightly integrated that to remove one part from its context is to render both the part and the system incomprehensible. The organism metaphor implies that the proper method for studying nature is not reduction but holistic synthesis: the goals of the investigator are to integrate and synthesize the scattered details of experience into a whole and to give some account of the unifying principles that connect everything together. The method tends to be speculative, on the assumption that the unifying principles in question can only be grasped by reason. The natural world as experienced through the senses and as studied by the natural sciences is a chaos of distracting details; the mind can only bring this chaos to order by the firm and consistent application of rational principles.

Advocates of the organicist worldview consider it to be more "ecological" than the alternative because it provides a more coherent—and more limiting—context for human activity. One feature that distinguishes life from non-life is that living things engage in goal-directed activity. Because organisms have goals and interests of their own, they can be harmed. If nature as a whole is really a kind of organic unity, then it has ends and interests of its own; nature can be harmed. Many environmental philosophers pick up on this implication, hoping that it might serve as a guide for human behavior. If human desires and projects can be brought into accord with nature's interests or nature's demands, they believe, then we may find a way out of the environmental crisis.

These advocates do not always make it clear whether the organicist worldview is supposed to be taken literally. There are at least a few environmental thinkers, Aldo Leopold among them, who openly acknowledge that arguments on behalf of organicism and holism are not so much true as useful. They maintain, in effect, that an ecological worldview should be embraced if and only if it supports the values and objectives of the environmental movement. On the other hand, a number of environmental philosophers do seem genuinely convinced that the usefulness of organicism must reside in its literal truth. I will consider this problem in some detail in Chapter 5; for the time being, I will assume that claims on behalf of organicism are meant to be taken at face value.

There are several variations on the organicist theme. A number of environmental philosophers, including some deep ecologists, environmental ethicists, bioregionalists, and others, focus in particular on the balance of natural systems and the interdependence of their component parts. Aldo Leopold's land ethic, published in 1949 as the conclusion to *A Sand County Almanac,* is the archetype for much of the philosophical work that has been done along these lines. The ethic was based on Leopold's understanding that energy and matter flow through natural systems, much like electricity flows through a circuit. The circuitry of nature, he believed, displays an orderliness and a balance that ought to be respected and preserved. In particular, Leopold valorized the complexity, integrity, stability, and beauty of biotic communities. This is a relatively static view of ecological systems: energy flows, but the circuits themselves do not change unless humans change them— usually for the worse.[3]

Others have developed a more dynamic organicism, drawing on their own interpretations of modern scientific cosmology and evolutionary theory. These include some deep ecologists, some social ecologists, and a few scientific and religious mystics who tell what they call "the universe story."[4] Dynamic organicism portrays nature as an unfolding reality, perhaps guided by a single, fundamental creative impulse; it places emphasis not so much on the stability of nature as on the broader movement of cosmic evolution. Advocates of the dynamic view might agree with Leopold that natural entities and systems are interdependent, bound together by a harmonious living circuitry. However, they would insist that the apparent equilibrium of the present moment takes its place within a much larger story in which the universe becomes more differentiated over time, like the branching of a great tree. Everything in the universe is connected to everything else by common ancestry.

Some environmental thinkers, ecofeminists in particular, have objected to the tenor of both variants of what might be called "strong" organicism, at least in part because of what are taken to be its unsavory political consequences. In effect, these critics accuse deep ecologists and their kin of presenting a false dilemma: either nature is a collection of resources or else it is a monolithic integrated whole in which the individual is to immerse herself or himself. The "oceanic" feeling of fusion overwhelms and threatens to obliterate the individual in what at least one critic calls an "androgynous" natural unity. The alternative to this heavy-handed metaphysics is a more modest sort of organicism, one characterized by a "feminine" emphasis on the web-like relations in which the individual is personally involved and which are "definitive of self."[5]

The question remains of how this ecological world-view is to be established. Broadly speaking, there are two options. Environmental philosophers could follow the pattern of speculative philosophy and attempt to gain a direct intellectual grasp of the unity of nature or they could search for the unity of nature in the results of scientific research, especially research in ecology and evolutionary biology.

The core of the speculative approach is the belief, or the hope, that the basic principles underlying the order of nature are simply self-evident. Plato believed that mind and nature are united by a common rationality, so that clear thinking must necessarily cut nature at its joints. Descartes called upon God to assure that his clear and distinct ideas corresponded to external reality, while John Locke and his fellow empiricists turned to the senses as the foundation for a coherent picture of the world. However philosophers try to attain certainty, their efforts are subject to the same basic requirements: if speculative philosophy is to work, there must be some sort of bridge between mind and nature so that one may be brought into conformity with the other.

Deep ecologists use a method of identification, which begins with the experience of having something in common with other people. When I identify with other people, when I recognize all that I have in common with them as experiencing subjects, the boundaries between my Self and their Selves are weakened to the point that I can no longer conceive of myself apart from my relationships with them. Devall and Sessions extend the same kind of identification to the natural world in a process they call "Self-realization." The individual Self identifies not only with other people but also with natural entities and systems, until the individual self ultimately identifies itself with the all-encompassing Self of "organic wholeness." The Self is nothing other than the relational field, in which everything is connected with everything else.[6] From this higher perspective, it is no longer possible to regard a tree, for example, simply as potential firewood. Because the tree participates in the relational field, each individual Self is bound together with the tree by an internal relationship, and neither can be conceived of as separate from the other.

Another deep ecologist, Warwick Fox, has made this account more complex by distinguishing three varieties of identification. In addition to personal experiences of affinity with other beings, on which Naess, Devall, and Sessions rely, there are also what he calls ontological and cosmological forms of identification. In other words, experiences of commonality are also brought about through a "deep-seated realization," first, of "the fact *that* things are," and second, of "the fact that we and all other entities are aspects of a single unfolding reality." Cosmological identification is made possible by any cosmology "that sees the world as a single unfolding process," that calls for "the empathic incorporation of mythological, religious, speculative, philosophical or scientific cosmologies." Science is the only source of cosmology in the modern world, and it provides insight into the broader context of human life: scientific inquiry can be interpreted as producing "an account of creation that is the equal of any mythological, religious, or speculative philosophical account in terms of scale, grandeur, and richness of detail."[7]

A further variation seeks a basis for identification

in the structures of perception itself, drawing from the tradition of phenomenology established in the early twentieth century by Edmund Husserl and developed most prominently by Martin Heidegger and Maurice Merleau-Ponty. The impetus for phenomenology is a critique of scientism or objectivism, the belief that the world of the sciences is the only true world. From the scientific point of view, reality consists of discrete, measurable physical objects moving about in absolute space; anything else is held to be merely subjective. Phenomenologists argue that scientism comes about when people forget that the scientific world has been abstracted from the far richer lifeworld that is the locus of human interests and human activity. To adopt scientism is to impoverish the world and to bring about a crisis of culture, whereby the sciences lose their meaning for human life. The basic point of the phenomenological method is to set aside or "bracket off" all presuppositions about what is real and to restore the richness, variety, and meaning of the world as it is encountered in lived experience.

For some, this leads to a kind of "phenomenological ecology." The argument generally proceeds as follows. The lifeworld, as revealed through phenomenological investigation, is an intertwining network of significations; each of the things I perceive has meaning only within the context of the larger perceptual unity of which it is a part. So, in perception at least, everything is connected to everything else. Phenomenology is rooted in an affirmation of the reality of the perceived, the belief that the lifeworld is not a subjective illusion that masks the objectivity of the physical objects that are the concern of science. So if the lifeworld is taken to be the primary reality of human experience, and if what I encounter in the lifeworld are relationships rather than detached physical bodies, then I seem to be entitled to adopt the principle that reality is essentially relational.[8]

There are still other variations on this theme, but I have focused on those that rely most explicitly on the speculative or intuitive method for discovering the unity of nature. Now it is time to bring to bear the principles of skepticism. Can the methods of speculative philosophy give environmental philosophers what they want? Do environmental philosophers who adopt these methods proceed with due modesty and caution? I believe the answer to both questions is "no."

HEGEL AND THE LIMITS OF PHILOSOPHY

For Plato, the cosmos is intelligible to the human mind because both are patterned on the same rational principles. To gain knowledge, it is necessary only to bring the mind into harmony with the universe, to make distinctions in thought that copy distinctions in reality. If there is any difficulty in finding the truth of things, it is because there is another causal principle, necessity or materiality, which operates independently of the rational cause and distorts its appearance in the world. Here, then, is the basic challenge for speculative philosophy: to filter out all of the distortions and details of the world of appearance in order to grasp the underlying rational unity.

This challenge may well exceed the capabilities of the human mind. A skeptical reading of the history of speculative philosophy gives rise to the suspicion that whatever method we use, however carefully we proceed, the universe will ultimately evade our efforts to grasp it. These doubts extend also to the various proposals for an ecological worldview, at least insofar as they rest on a speculative foundation. If the finest, most rigorous philosophers in the Western tradition have all fallen short of the goal, has speculative philosophy progressed so far in its methods or principles that environmental thinkers might now succeed? This strikes me as unlikely.

I will take up the story in the early modern period, just after Descartes. Speculative philosophy in this period differed from what had come before, largely because Descartes introduced a representational model of consciousness. In the dualist universe, the mind is a substance that thinks, which is to say that it holds within itself a play of ideas. Ideas are like pictures held before the mind's eye, disembodied reflections of things both real and imagined. Knowledge, in this view, consists of a correspondence between ideas and the real things they are supposed to represent. The rationalists appealed to God to guarantee correspondence, while the empiricists appealed to the senses. Hume's skepticism is informed by the difficulties encountered by rationalists and empiricists alike and can be read as his recognition of the limits of the representational model of consciousness.

There has been quite a lot of speculative philosophy since Hume's time, of course, much of it based on modification or outright rejection of representationalism. One particularly fruitful tradition passes through Immanuel Kant and Romanticism to the nature philosophy of Hegel and concerns the possibility of unifying mechanistic and organic accounts of nature. This is particularly interesting for my purposes. Deep ecologists and others explicitly claim Romanticism as one of their sources of inspiration, and Hegel's attempt to construct a coherent, speculative nature philosophy stands as one of the most rigorous and systematic in the philosophical tradition. If he fails, others would be hard pressed to succeed.

Nature, Hegel writes, "confronts us as a riddle and a problem, whose solution both attracts us and repels us:

attracts us, because spirit is presaged in Nature; repels us because Nature seems an alien existence, in which spirit does not find itself."[9] When I consider the world under the aspect of internality, each thing and event can be understood in terms of a larger whole. The tree outside the window and my own body share an irreducible, internal relationship to one another; each is part of the living order of nature. The human mind is drawn into the web of internal relationships because it recognizes itself there. Order responds to order. When I consider the world under the aspect of externality, however, things and events seem disconnected from me and from one another. My body and the tree outside my window have nothing essential in common. It just happens that both are material and both exist in space at the same time. The human mind can find no foothold in the shifting details of accidental material relationships. There is no order to which it can respond.

The unresolved contradiction of nature consists in the fact that the mind is caught between these two interpretations. Nature is at once familiar and strange, benevolent and hostile—or perhaps simply indifferent. Hegel insists that if there is to be any solution to the riddle of nature it can only be reached through a systematic internalization of the external world. This means that the mind must seek out and articulate the basic rational unity of nature in such a way that no detail remains unconnected or unexplained.

Hegel can be read in various ways, and there is a good deal of debate over how seriously to take *The Philosophy of Nature*. It may be that Hegel introduces the riddle of nature because he intends to solve it once and for all. If so, then he promises nothing less than absolute and certain knowledge of the natural order. On the other hand, he may introduce the riddle because it has no solution. In that case, it is just an inescapably difficult moment in the development of human consciousness. But my interest in Hegel is hypothetical: if Hegel is seriously trying to solve the riddle of nature, then his success or failure will reflect on the efforts of speculative environmental philosophers to solve the riddle for themselves. For the sake of argument, then, I will assume that Hegel's nature philosophy is a careful systematic effort to attain absolute knowledge of nature.

First, though, I should provide some historical background to Hegel's work. The distinction between internality and externality I have just outlined is, in effect, a modern recasting of Plato's distinction between reason and necessity. In the wake of Cartesian dualism, all life, purpose, and value—all final causes—came to be attributed to the mental realm. Matter, as I have discussed, was thought to be inert, its only attributes being those of taking up space and of being infinitely divisible. For Descartes, everything in nature is external to every other thing and, because it is divisible, matter can always be made external to itself. If I break a stone in half, I am left with two stones which no longer bear any relation to each other aside from their proximity in space and the possibility that I could hit them together. If nature has any order or purpose, it is only the efficient purposiveness of a machine, the organizing principle of which resides in the rational mind of the machinist-god who created it and set it running. Even the human body is a machine, albeit a wondrous one; Gottfried Leibniz, a philosopher who followed and extended Descartes' rationalism, described the body as being so subtle that even its parts are themselves machines.[10]

This account of the origins of mechanism has the surprising consequence that mechanism and organicism actually have a lot in common. Insofar as they are speculative principles, both appeal to rationality as the model on which the natural world is patterned, and both emphasize the quintessentially rational capacity to set goals. The difference between organicism and mechanism comes down to the difference between internality and externality, which has consequences for where the organizing principle of nature is thought to reside. The development of an organism is interpreted as a process of striving to fulfill its own specific form, an organizing principle that resides within the organism itself and subordinates every part to the life of the whole. The orderliness of a machine, on the other hand, can be explained only in terms of the conscious intentions of the machinist, who shaped the parts and arranged them for some purpose or other. So to interpret the cosmos as an organism is to place the rational organizing principle within the cosmos itself, while to interpret the cosmos as a machine places the organizing principle in the mind of a supernatural creator.

The Romantic effort to internalize externality can be understood only in light of Kant's critical philosophy, which served as both predecessor to and foil for the Romantic imagination. Kant came to see early modern philosophy as a dead end: the rationalists' exclusive reliance on pure principles of reason without appeal to the senses rendered their systems empty and dogmatic, while the empiricists' exclusive reliance on sensory experience without appeal to pure rational principles rendered them blind. Kant sought a third way, but in doing so he had to change the terms of the debate. In essence, he turned his focus inward, away from the world and the hope for correspondence to the character and limits of human cognition.

From his investigation of the understanding, set out

in the *Critique of Pure Reason,* Kant drew the conclusion that experience and knowledge of nature are possible only if some pure or rational element is brought to bear on the chaotic flux of sensation, organizing it into the world we experience and establishing the foundations of science. For example, Kant held that the idea of causality comes neither from the senses alone nor from pure reason alone. It is possible to know causal laws of nature, he insisted, because causality is a pure category of the understanding that is brought to bear on sensory perception. The human mind has the built-in capacity for making hypothetical assertions of the form "if x then y," which serves as the template for any causal law of nature. Sensation can provide only the material, the x and y that are plugged into the formula. All told, there are twelve pure categories of the understanding, and taken together they provide the foundation for the science of nature. The natural world as it is grasped by the understanding is very much as Isaac Newton envisioned it, consisting of material objects moving through space and governed in their (external) relations to one another by deterministic causal laws. As far as the understanding is concerned, every phenomenon can and must be explained in these terms.

This is not merely an endorsement of mechanistic metaphysics, however. By changing the terms of the debate over knowledge, Kant turned his back on any possibility of a direct link between mind and reality. Nature, as he understood it, is no longer the totality of things as they really are, but a construct of the understanding. Technically, he defined nature as the possibility of future experience, but he saw experience itself as generated by the understanding as it imposes the order of the categories on the chaos of sensation. I can have no knowledge of things in themselves, Kant insisted, only of things as they are for me. The product of the understanding is called knowledge, but not because it corresponds with what is really out there. Rather, human knowledge of nature is valid to the extent that it is consistent: the categories of the understanding are universally valid, so that any rational being would constitute its experience in precisely the same way. Nevertheless, because he construed the understanding as only one of the faculties of the human intellect, and because he denied that the understanding has any access to things as they really are, Kant left open the possibility that there are other, legitimate ways of thinking about the natural world and its relationship to human beings. This is important, because it made possible the reconciliation of opposites that had long been considered irreconcilable.

Consider Kant's treatment of final and efficient causality in *The Critique of Judgment*. In its reconstruction of experience, the understanding encounters entities that are beyond its grasp. The material causal connections within an organism, for example, interact reciprocally in such complex ways that the connections can only be judged to have been brought about for some purpose. The physiology of the human body, for example, only makes sense if it is the product of the conscious, free act of a rational being rather than the product of blind determinism. Unlike a watch, which has only a motive force based on the external relations among the parts, an organism is best thought of as possessing a "formative" or developmental force based on the internal relations among the parts. So, contrary to the understanding, we judge an organism to be "both cause and effect of itself." The eye is a condition for vision, but I can also judge that the eye exists for the sake of seeing, as though the idea of sight preceded the existence of the eye in the mind of some designer.

Kant calls such entities "natural purposes," which for a Cartesian would be a contradiction in terms. While he acknowledged that it would be impossible to give a scientific account of the coexistence of efficient and final causality in natural purposes, he insisted that it is not at all contradictory for the judgment to posit their coexistence. The two forms of causality do not contradict each other because they arise from different faculties of cognition: linear efficient causality grasped by the understanding is "real" while final causality posited by judgment is "ideal." Because I have no access to things in themselves, Kant believed, the best I can do is to follow the categories of the understanding as far as possible and to account for the rest as reason and judgment see fit.[11]

Following the pattern established with individual organisms, Kant argued for the possibility of judging nature as a whole to be a unified purposive system—one that has come into existence for the sake of supporting humanity as a community of free rational beings. Kant judged humanity to be the final purpose of creation because humans are the only earthly beings capable of setting purposes for themselves. "Only in man," he wrote, "and even in him only as a moral subject, do we find unconditioned legislation regarding purposes. It is this legislation, therefore, which alone enables man to be a final purpose to which all of nature is teleologically subordinated."[12] From this basis Kant formulated a "moral theology" in which God is the thinking and willing agent who has created the natural order in such a way that it harmonizes with and supports human culture. Hence, the unity of freedom and nature is established as a subjective principle of reflective judgment. This is part of his answer to the question "For what may I hope?" If I can judge nature to be a teleological system and its goal to support humans in their vocation as moral beings, then I can at least hope that good deeds will be rewarded, and bad deeds punished, by the system of nature itself.

Romanticism emerged in the early nineteenth century as a broad cultural and intellectual movement of reaction against modernism in philosophy and science, guided by the vision of a threefold unity: the unity of knowledge, the unity of nature, and the unity of human consciousness, or spirit, with nature. Among the literary, artistic, and religious expressions of the general Romantic movement, *Naturphilosophie* stands out as that form of Romanticism that seeks to arrive at this threefold unity by way of the study of the natural world.[13] Schelling, one of the leaders of the movement, declared that "Nature should be Mind made visible, Mind the invisible Nature. Here then, in the absolute identity of Mind *in us* and Nature *outside us*, the problem of the possibility of a Nature external to us must be resolved."[14] With this assertion, Schelling took up Kant's conclusions regarding the reconciliation of the natural and moral realms, but he and his compatriots cast aside Kant's caution by asserting the real truth of the reconciliation. For the Romantics, nature *really is* the self-externalization of mind, and the individual human mind is a product of nature that serves as the means by which nature awakens and becomes conscious of itself as spirit. This is also a departure from Kant in that knowledge of nature is to be formulated in terms of a dynamic process of development analogous to that of an organism rather than in terms of the static categories of traditional logic.

This brings my account back to Hegel, who gave this dynamic interpretation of nature its most powerful and sophisticated expression. This is not to say he was in full agreement with his contemporaries. In fact, Hegel made a point of distancing himself from Schelling and the other Romantics. In their "unskilled" hands, he asserted, nature philosophy had received "crude treatment" and so had been "brought low not so much by its opponents as by its friends." He identified the problem as simply laziness or, at best, sloppiness. The Romantics proceeded

> not by bringing consciousness out of its chaos back to an order based on thought, nor to the simplicity of the Notion, but rather by running together what thought has put asunder, by suppressing the differentiations of the Notion and restoring the *feeling* of essential being.

The result was not a coherent philosophical system but simply a kind of "rapturous haziness"; it lacked "the seriousness, the suffering, the patience and the labor of the negative."[15]

The "labor of the negative" is one way to describe dialectical logic, which Hegel developed as a logic for comprehending dynamic processes. The Concept (i.e., universality as such) begins as a simple unity; this is the first moment of the dialectic. When this unity is negated, it falls apart into a contradiction. In this, the second moment, consciousness vacillates between two interpretations of itself that are incompatible. The contradiction can only be overcome when consciousness recognizes that the two terms of the contradiction have something in common, which serves as the basis of a new unity. Hegel called this third moment "determinate negation": the contradiction is negated, but in such a way that the distinction it establishes is preserved as part of the new unity. As a result, the Concept becomes more comprehensive and less abstract; it becomes a unity-in-difference.[16] Ultimately, the new unity itself is negated, and the process begins again. To engage in the "labor of the negative" is to work through every step of the logic as carefully as possible, noting every distinction and every resolution, making sure that each step follows by logical necessity from the one before. The end result of this labor is "absolute knowing," in which the universal and the particular are unified. When the Concept has already comprehended every possible concrete distinction, reason can no longer extend itself because nothing remains that is other to it.

In Hegelian thought, the problem of nature arises as the second moment of an overarching triad. The Concept begins in unity and then falls apart into particularity and loses itself in the contingency and detail of externality. This is the unresolved contradiction, the riddle of nature. With fractal complexity, nature itself develops dialectically: each moment in each triad resolves into yet another triad. Mechanics gives way to physics, which gives way to organics; within organics, geological nature gives way to plant nature, which gives way to animal nature; animal nature resolves into a triad of its own, and so on. Overall, the process moves away from externality and the rule of efficient causality toward internality, subjectivity, and spontaneity, all of which paves the way for the development of human culture. At the end of the process, the fractured Concept comes back to itself and becomes aware of itself as spirit. What is to count as knowledge of nature is this "path of return . . . which overcomes the division between Nature and Spirit and assures to Spirit the knowledge of its essence in Nature."[17]

Organic nature is, for Hegel, the external expression of subjectivity. An organism has its existence in the external world, and yet it has the same kind of internality that characterizes human consciousness: it is both cause and effect of itself. As organic nature develops, this internality or subjectivity becomes stronger and more self-assertive. In the first moment, geological nature, the Earth as a whole is taken to be a kind of proto-organism, but one in which internality is still dominated by externality. In other words, the Earth-organism has the abstract shape of internality, but it does not have life of its own because it does not yet have the concept of itself. Plant nature is a significant improvement, possessing a "truer vitality," but it is still unable to express itself as a unity. An individual plant continually "falls apart" into modular parts that remain outside and relatively independent of one another; this is why it is possible to propagate a plant through cuttings or to graft part of one plant onto another. Animal life brings together the members of the organism into an ideal system of subjectivity; that is, into a self-sufficient organic unity. It is only with the emergence of the human animal, however, that the Concept is finally able to overcome the contingency of externality by attaining consciousness of its own essence as spirit. To paraphrase: human culture is a higher kind of order, one which is able to maintain its own integrity even in the face of the death of any individual human. More to the point, human culture can finally grasp that the purpose of nature, the goal of its entire development, is to produce human culture.

Many environmental thinkers would dismiss Hegel's nature philosophy out of hand because it is explicitly human centered. Even so, Hegel's conclusion is consistent with the historical pattern of speculative philosophy. To the extent that environmental philosophers engage in speculative nature philosophy, they must be aware of the pervasive tendency to anthropomorphize the natural order. For my purposes, however, the important question is this: Has Hegel succeeded in comprehending nature; has he really grasped the essence of things? If he has not, it will be helpful to see where he has fallen short and to consider the consequences for more recent efforts along these lines.

Scattered throughout the *Philosophy of Nature* are what can only be called "disclaimers" on behalf of speculative philosophy. Hegel repeatedly insists that it is not his responsibility to account for every detail of nature, citing either the difficulty of philosophy when confronted by externality or the "impotence" of the Concept in nature. In any case, it is clear that Hegel is primarily concerned with using nature as a stepping-stone to spirit. The *zusatz* of the final paragraph includes a particularly revealing statement:

> The difficulty of the Philosophy of Nature lies just in this: first, because the material element is so refractory towards the unity of the Notion, and, secondly, because spirit has to deal with an ever-increasing wealth of detail. None the less, Reason must have confidence in itself, confidence that in Nature the Notion speaks to the Notion and that the veritable form of the Notion which lies concealed beneath Nature's scattered and infinitely many shapes, will reveal itself to Reason.[18]

At first glance, it seems odd that Hegel should have to encourage reason to have "confidence" in itself "none the less"; that is, in spite of its own impotence in the face of externality. Consider the language he uses when he heralds the emergence of spirit from the death of nature: *"from this dead husk,* proceeds a more beautiful Nature, *spirit."*[19] What is this "dead husk" and what happens to it afterward? In one of the final passages of the *Philosophy of Nature,* Hegel recognizes that nature philosophy is difficult because the material element is "refractory" toward the unity of the Concept such that the Concept is "impotent" to hold itself together. This seems to imply that there is *something* "out there" that is *other* to the Concept; the Concept would not distort or conceal itself. The conclusion seems unavoidable that the materiality and diversity of the external world are *other* to reason and that Hegel can proceed only by actively ignoring them. It would seem that, since Plato's time, as the rational cause has been refined and more clearly grasped, necessity too has been refined so that it more completely evades reason's grasp. In Hegel, it would seem, necessity, or externality, has its ultimate revenge, receding to its place as the ultimate and intractable other of reason, the unintelligible core of nature.

Jacques Derrida offers a helpful metaphor that sheds some light on the problem: Hegel's system of speculative philosophy is a "restricted" economy, one that is

> limited to the meaning and to the established value of objects, and to their *circulation*. The *circularity* of absolute knowledge could dominate, could comprehend only this circulation, only the *circuit of reproductive consumption*. The absolute production and destruction of value, the exceeding energy as such . . . all this escapes phenomenology as restricted economy.

With his determinate negation, Hegel hoped to negate and to conserve at the same time, while he ignored and externalized the possibility of an absolute negation, the "*indefinite* destruction of value."[20] Beyond the tidy circle of this economy, in which meaning is so carefully conserved, is "the exceeding energy," or simply "excess," the non-meaning out of which meaning is produced, into which it returns, and through which it is destroyed. Hegel must externalize excess because it is utterly other to reason and so cannot be brought into account. Even so, with his casual references to the dead husk and to the refractoriness of the material element, Hegel occasionally allowed externality to show up on the books, if only for a brief moment.

The endless details of nature open up the possibility that new discoveries about those details could render obsolete all of Hegel's carefully reasoned accounts of natural processes. After all, speculative nature philosophy has no self-correcting mechanism with which to keep pace with new and surprising phenomena as they emerge. Simply as a matter of historical observation, many of Hegel's accounts are simply wrong, often laughably so, in spite of his insistence that they are logically necessary. According to Hegel's speculative geology, for example, the geological Earth stands at the threshold of life: it embodies subjectivity, but only in an abstract form. Out of this diffuse sort of internality, "every stone breaks forth" into a profusion of "punctiform" proto-life; this is his account of the origin of lichens and mosses. The Earth also makes other "playful essays in organic formation," such as the shapes of trees that are sometimes found in coal seams. Hegel insists that these shapes could not be the remains of living things that have died but are rather "stillborn" organic formations that can be expected to occur on the cusp between two moments in the conceptual development of organic nature.[21] Hegel's denial of the possibility of fossils is grounded in his rejection of contingency in nature. The appeal to mere time, he insisted, does not account for the structure of the world, because all change in nature is to be understood as *logical* rather than merely mechanical and temporal.

Within a quarter-century of the completion of the *Philosophy of Nature,* Charles Darwin revolutionized natural history with the publication of his *Origin of Species.* Hegel, had he lived, could have responded in one of two ways. On the one hand, he could have ignored Darwin on the principle that scientists can only stumble around in the external world, chasing after details, while reason grasps at the essence of things. The consequence would be that the passage of time and the ongoing growth of the sciences would render Hegel's speculative nature philosophy obsolete and increasingly irrelevant to human affairs. The case of Darwin is especially troublesome for Hegelian thought: Darwin, more than any other, gave contingency and temporality a central role in explaining the apparent order of nature.

On the other hand, Hegel could have changed his own "logically necessary" accounts to integrate scientific theories of his contemporary. In response to Darwin, Hegel could have taken up and insisted on the necessity of evolution as a basic principle of nature, as many more recent speculative philosophers have done. At the very least, the idea of updating nature philosophy in order to keep up with the fruits of scientific "stumbling" is embarrassing for speculative reason. Worse, Hegel would probably have had to admit that the apparent strength of the sciences lies precisely in what he had identified as their weakness: the sciences have a self-correcting mechanism that allows scientists to adapt to new circumstances, while speculative reason is stuck with its own mistakes. The worst of it is that the appropriation of scientific theories for speculative purposes can only be carried out by misconstruing the scope and limits of scientific inquiry and the degree to which scientific concepts and theories may legitimately be transferred from one intellectual domain to another. This is a pervasive problem in speculative environmental philosophy.

If, according to the first principle of skepticism, my thinking is to be accompanied by doubt and caution, then

Hegel's admonishment that reason should have faith in itself seems to be a bad bargain. As far as I know, the externality that evades reason continues to hide the springs and principles of nature, just as it hid from Hegel the basic insights of evolutionary biology. If, according to the second principle of skepticism, my inquiries should be confined to those that are suited to my limits, then it seems that I am compelled to abandon the hope that nature is ultimately intelligible to reason. Hegel's own difficulties suggest that belief in intelligibility can be sustained only by actively ignoring anything that is unintelligible; in effect, I must presuppose intelligibility in order to discover it.

Selective Philosophy

The historical argument I have just completed has its limits; at most, it can introduce some reasons to doubt that speculative nature philosophy can succeed. Even though these reasons are serious, even though it is difficult to conceive of a speculative method that can overcome the problem of externality, the historical argument alone cannot establish once and for all that success is impossible. To get back to the thread of the argument, there is still the open question of whether environmental philosophers might overcome the limits of the tradition and actually grasp the nature of nature. It is perfectly consistent with the principles of skepticism to admit that this might yet happen. A closer look at the actual work of environmental philosophers, however, reveals further grounds for doubt.

In the introduction, I distinguished two interrelated projects of environmental thought: the radical project of finding and rooting out the intellectual cause of environmental destruction and the speculative project of establishing a more benign ecological worldview. As it happens, the two projects may not be so easily separated as the distinction implies. Indeed, it seems as though the speculative project is itself essentially reactive: environmental philosophers support organicism not because they have achieved some sort of privileged insight into the nature of things but because they are convinced of the vital political necessity of opposing mechanism and its consequences. As a consequence, environmental philosophers tend to work backward from the conclusions they would like to support to the first principles that strike them as best suited to the task of supporting those conclusions.

This is most obvious in the practice that I call "selective philosophy," in which environmental thinkers sift through philosophical and literary traditions, seeking out and exploiting any resource that might lend some weight of authority to the idea that reality is essentially relational. For example, Bill Devall and George Sessions report that in addition to ecology and physics, deep ecologists draw inspiration from the perennial philosophy (including that of Baruch Spinoza), Romanticism, literary naturalism and pastoralism, St. Francis of Assisi, feminism, the worldviews of primal peoples, Martin Heidegger, Taoism and other "Eastern spiritual process traditions," poet Robinson Jeffers, and John Muir.[22] Their comments on the perennial philosophy are particularly telling:

> An appropriate metaphysics for the emerging ecological perennial philosophy would provide a structural account of the basic unity and interrelatedness of the universe, while at the same time accounting for the importance and uniqueness of individual beings. Similarly, this metaphysics of interrelatedness helps us realize that the natural world and other species are inextricably a part of us, and us of them (a mutual reciprocity).[23]

The standard of what constitutes an "appropriate" meta-physics has been established in advance, and it has little to do with the truth of its first principles: a system of metaphysics is appropriate if it supports the platform of the deep ecology movement.

It is a common practice for philosophers who wish to involve themselves in environmental affairs to write apologies on behalf of one or another historical figure, defending his or her record of ecological thinking and claiming him or her for the environmentalist cause. The range of possible sources continues to expand; it includes Gottfried Leibniz, Karl Marx, Martin Buber, John Dewey, C. S. Peirce, Maurice Merleau-Ponty, Friedrich Nietzsche, and anyone else whose work can be recast in an environmentalist mold. Taken together, the resulting list of sources can only be counted as bizarre, lending credence to the saying that politics makes strange bedfellows. Spinoza and Nietzsche, let alone Heidegger and St. Francis, would hardly see eye to eye on the nature of nature. An alliance among feminists, Taoists, and primal peoples seems, to say the least, a bit forced.

The problem of selective philosophy is complicated by the fact that the term 'nature' is itself contested territory, subject to any number of interpretations. My own argument concerning the limits of speculative philosophy, for example, assumed that nature is the system of laws or principles that binds together everything that exists. In one version of this account, even the ugliest of cities and the most destructive of weapons are perfectly natural simply because they exist and obey the laws of physics. However, it is also possible to conceive of nature as radically other to human culture, as that which is untamed, uncontrollable, and unintelligible at the boundaries of human experience. Wild nature, in this sense, is very much like what Derrida calls "excess," that from which all meaning springs but which itself threatens the destruction of meaning. Wild nature is a threat to human aspirations simply because it is profoundly indifferent to those aspirations.[24]

Both of these accounts are unsatisfying to those environmental philosophers who seek in nature some basis for judging and limiting human action. They need to maintain both that nature is something that stands over against culture and that there is something valuable—or at least approachable—in it. The solution, for many, seems to be to conflate the organicist notion of the system of nature with the idea of wilderness, that which is untouched by human hands. Wild nature, they argue, is the order of things as it was meant to be, before humans started messing things up; it is the standard against which all human-induced alterations are to be measured. The ideal of wild nature is also informed by aesthetic judgments and personal desires regarding certain landscapes, especially when primal forests or wetlands, for example, are contrasted with patterns of human development. In short, the environmentalist conception of nature has been constructed out of diverse and perhaps contradictory elements.

Like the Romantics before them, many environmental thinkers hold that it is at least possible for humans to identify with wild nature and to reintegrate themselves into its order. Aldo Leopold, for example, hoped to restore humans to plain citizenship in biotic communities, a hope that reverberates through the work of many later environmentalists. The most radical among them see the human-nature reunion as necessitating an abandonment of civilization.[25] Reintegrating humans into wild nature, as it has been defined by environmentalists, introduces an apparent contradiction; it is not clear that humans can do this while retaining their humanity. More traditional environmental ethicists, by contrast, simply wish to establish a new relationship between humanity and wild nature, between self and other, that is based on respect rather than exploitation.

Nature is everyone's favorite weapon: it is common practice to label something as "natural" in order to establish its value beyond all dispute. In defense of the inequities of capitalism, for example, social Darwinists used their own selective reading of evolutionary theory to argue that cutthroat competition is part of nature's plan.[26] Religious and social conservatives oppose homosexuality on the grounds that it is "unnatural," that it goes against God's plan of creation. The environmentalist conception of wild nature has also emerged out of a struggle for the hearts and minds of the public, constructed at least as much for its usefulness as for its truth. At the beginning of the twentieth century, John Muir and his allies fought to preserve the landscapes they loved from the encroachment of development. In effect, two visions of wilderness were in conflict. When Europeans came to the Americas, they brought with them axes, plows, and a view of wilderness as both a threat to humanity and a waste of good farmland. Muir argued, to the contrary, that wilderness is itself sacred and that it should be preserved as a refuge from the ugliness of civilization.[27]

From a skeptical point of view, the various constructions of 'nature' make any appeal to the term highly suspect. The same goes for 'environment,' which has been subject to the same kinds of reconstruction. Rather than referring simply to the surrounding world with which an organism interacts, 'environment' has begun to serve as a stand-in for 'nature', encompassing some of the same contradictory elements. Both terms should be approached with caution. It might even be a good idea for environmentalists to abandon the term 'nature' altogether and to limit their use of 'environment' to something like its original meaning. The only question, then, is whether anything remains for environmental philosophers to contribute to broader efforts to cope with environmental change.

[2]

Organism and Mechanism

Given the shortcomings of speculation, it is no surprise
that many environmental philosophers turn to the natural sciences to bolster claims on behalf of relatedness. It is
also no surprise that ecology is the most common source
of inspiration, although evolutionary biology and quantum mechanics have both served this function as well.
Insofar as it concerns the interactions among organisms,
many argue, ecology tends to break down the barriers between what have traditionally been regarded as independent entities. Bill Devall and George Sessions have claimed
that ecology has "provided a view of Nature that was lacking in the discrete, reductionist approach to Nature of
the other sciences," and so it has helped bring about the
"rediscovery within the modern scientific context that
everything is connected to everything else." Likewise,
J. Baird Callicott, one of the foremost interpreters of the

implications of the sciences for environmental ethics, has claimed that recent developments in both ecology and physics make it "impossible to conceive of organisms . . . apart from the field, the matrix of which they are modes." The matrix to which he refers is a system of internal relations that constitutes a "structured, differentiated whole."[1]

It is not difficult to understand the appeal of ecological concepts. Consider one possible line of argument. Some ecologists have come to believe that ecosystems— not organisms—are the basic units of nature. Suppose they were to demonstrate that ecosystems are nevertheless similar to organisms, at least in some respects: they maintain homeostasis, for example. Suppose further that an environmental philosopher were to pick up on this similarity and draw from it the implication that ecosystems are like organisms in morally significant respects: they have an interest in maintaining homeostasis, which implies that they can be harmed. If they can be harmed, then it might follow that humans have an obligation to prevent harm to ecosystems or even to repair any such harm once it has been done.

The argument of this hypothetical philosopher has much in common with the sort of speculative organicism I discussed in the last chapter, especially in the claim that ecosystems exhibit traits associated with internality or subjectivity. The difference is that ecological theory— with its rather narrow and cautious organicism—serves as a vital link in this chain of reasoning. As scientific inquiry has effectively become the final arbiter of knowledge concerning the natural world, many would regard support from ecology as decisive. A consistent skeptic, however, would not take the authority of the natural sciences at face value, especially when that authority is borrowed by non-scientists for political purposes. In this case, the key question is this: May environmental philosophers legitimately use the findings of ecologists to inform much broader claims on behalf of organicism?

While there is widespread agreement among environmental philosophers that ecology can support an organic interpretation of nature, there is actually quite a lot of disagreement among them about the details of that interpretation. Callicott, for example, claims that deep ecologists are mistaken in their belief that the "organic wholeness" with which humans are to identify is a homogenous whole, one without internal ontological distinctions. The science of ecology, he argues, supports instead a picture of the world as a differentiated whole in which the individual is not entirely swallowed up by the network of which it is a part. Callicott regards his own ecocentric ethic as being more adequate precisely because it is more "ecological."[2]

Karen Warren and Jim Cheney, in their ecofeminist critique of Callicott's metaphysics, reject even this more sophisticated understanding of wholeness. Instead, they draw on their own understanding of hierarchy theory. A recent development in ecosystem ecology, hierarchy theory, is based on the insight that ecological systems can be studied at a number of different levels, each of which is relatively independent of the others, and each of which calls for its own methods and theories. So the relation of a given species to its environment will be studied in one way, while the more encompassing systems of global ecology will be studied in quite another, and each kind of research is valid in its own domain. Hierarchy theory yields the further insight that these various levels of organization may be arranged hierarchically, where each is nested within but relatively independent of the more encompassing levels. This leads Warren and Cheney to assert, contrary to both Callicott and the deep ecologists, that "overconnectedness in a system . . . is *un*stable," and that an "adequate metaphysical ecology" must recognize that "the world 'strives,' so to speak, to organize itself into discrete and relatively disconnected or autonomous holons and hierarchical levels of organization as a condition of its own stability."[3]

All of this debate passes over two much more important questions. First, is ecology really founded in holism and organicism, as many environmental thinkers claim? The answer is to be found in the history and philosophy of ecology, on the basis of which I will argue that ecology is neither unambiguously organicist nor unambiguously mechanistic. This leads to the other, broader question: To what degree may scientific concepts and theories legitimately be transported from one intellectual domain to another? The answer is to be found in the character of scientific knowledge itself and the demarcation between the sciences and other aspects of human thought. I will argue that even if ecology did embody a true scientific holism, it is not at all clear that environmental philosophers would be entitled to draw from it the kind of metaphysical or ethical implications they desire. From a skeptical point of view, an appeal to the natural sciences to support an ecological worldview is unlikely to meet with any more success than an appeal to speculation.

PHYSIOLOGY AND THE EMERGENCE OF ECOLOGY

Throughout the history of ecology, and within the context of current ecological research, mechanical and organic principles have acted as metaphors. The natural world is complex, and ecologists have appealed to more familiar—

and less complex—phenomena in order to model that complexity and make it comprehensible. In this, ecologists are no different from other scientists. Like their fellow scientists, they are not content simply to invoke metaphors and hope for the best. Precisely because the natural world is complex, simple models tend to break down over time; the point of scientific inquiry is to subject such models to critical scrutiny, to test them against experience, and to refine or reject them as necessary.

Through this process, ecologists have combined and refined the organic and mechanistic metaphors, pushing them toward greater and greater particularity and specialization. In their speculative context, the principles of organism and mechanism have the broadest possible application, since they are intended to account for the universe as a whole. In ecology, they have been used only to answer very specific questions about the interactions among living beings and between living beings and their physical surroundings. This places strict limits on the conclusions that can be drawn from ecological theory both within the sciences and in other domains.

When the first ecologists came on the scene at the end of the nineteenth century, they brought with them metaphors that had already been refined through earlier developments in biology. Once again, the story begins with Descartes. For Descartes and other early modern thinkers, living beings were nothing more than very complex and subtle machines. This purely mechanistic interpretation of organic structure was fashionable for about fifty years, but the accounts it generated proved to be dissatisfying, even absurd. Given the limits of both mechanics and physiology at the time, for example, there was no plausible way of accounting for complex living functions such as reproduction. In the eighteenth century, the challenge for those who studied living things was to get beyond simplistic mechanism while remaining firmly materialist; that is, without appealing to supernatural agencies.[4]

Some Romantic biologists settled on a vitalist theory —a theory of "living matter"—which held that there is an irreducibly living principle in terms of which plants and animals could be explained. This was not merely an exercise in speculation, however. In the middle of the nineteenth century, life scientists who were inspired by Romantic ideas began to distance themselves from the speculative excesses of *Naturphilosophie* in favor of a more rigorous experimental method. "Life" and "organic functioning," which had attained the status of logical categories in Hegel's system, could no longer be taken for granted but themselves needed to be investigated and explained. What characterized the work of these scientists as "Romantic" was their rejection of mechanistic metaphors as guiding concepts, even though they also rejected the elevation of organicist metaphors to unquestioned cosmological truth.[5]

Oddly, Romantic biology was, in its fundamental perspective, very much in tune with the decidedly unromantic work of Lamarck, who is credited with first using the term "biology" to designate a distinct scientific discipline. Lamarck, who worked as the eighteenth century changed to the nineteenth, opposed the vitalistic theories of "living matter" that abounded in France before the Revolution. Life, he maintained, was to be understood as the result of certain combinations of ordinary, "dead" matter. All the same, living things could not be accounted for in terms of the basic laws of physics, only in terms of their organization itself: "if on the one hand it is probable that life begins by a physical-chemical process . . . on the other hand it is quite clear that once that matter has formed itself into an organized body, it ceases to be a mere manifestation of mechanics."[6]

Even today, biology continues to develop through this fundamental tension between mechanism and organicism. On the one hand, living functioning is understood in terms of physical structures and chemical processes that can be analyzed into their various components. On the other hand, because biology is the investigation of *organized* beings, it passes beyond the point where mechanistic models can apply. The various organs and processes that make up a living being must also be understood in terms of the whole of which they are parts. François Dagognet, a contemporary philosopher of medicine and biology, balances organicism and mechanism in his statement of "the spirit of physiology":

> The living thing is assuredly a "whole," but, to know it as such, it is still necessary to disassemble it. Its unity does not come from identical and conglomerated elements, simply stacked together. They are assorted, correlated, different. In order to be assured of this, it is necessary to enter into the concert, to disarticulate, in one manner or another, without altering the idea of this whole. . . . [It is] the *exteriorization* of an *interiority* which, otherwise, if it had not been exteriorized, would be a "negative" interiority, which is to say unknown, obscure and confused; one would thus inhabit a night in which all cats are grey![7]

Ecology emerged as a distinct scientific discipline at the end of the nineteenth century, growing out of the long-standing traditions of natural history and natural theology. Early ecologists interpreted units of ecological organization as being very much like organisms, borrowing their models from both speculative philosophy and the newly refined science of physiology. This point is important enough to bear repetition: the starting point for modern ecology was not simply a sweeping and speculative organicism like that of the Romantic nature philosophers but the complex mixture of machine and organism that had emerged through more than a century of biological thought. However, because biotic communities and ecosystems are unlike organisms in many ways, it was inevitable that ecological theories would diverge from their speculative and physiological counterparts. To understand this divergence and its impact on ecology in the twentieth century, it will be helpful to contrast the work of an ecologist of the late nineteenth century, Stephen A. Forbes, with the work of Carl Linnaeus more than a century earlier and give some account of the ideas and influences that separate them.

Linnaeus, and those of his school, conceived of nature as a unified whole, the internal structure of which is arranged by God to maintain itself in a state of balance or harmony. Indeed, the evidence Linnaeus found for the harmony of the creation was meant to serve, in the spirit of natural theology, as evidence for the existence of a provident creator. In any case, Linnaeus used the metaphor of "economy" to describe the harmonious balance of nature, which he took to imply that each member of the natural community had a role to play in the life of the whole. This same metaphor had been used in "animal economy," a predecessor of physiology that attempted to characterize the structure of individual organisms in social terms.[8]

Mechanical, organic, and social metaphors overlap freely in Linnaeus's work on the economy of nature, bound together by their common emphasis on function, integration, and final causality. The mechanical metaphor entered in part because of the influence of Newton, Locke, and others and in part because of the simple fact that the creator of the economy of nature is not a part of the economy itself: the organizing principle of nature is outside nature, and natural theologians can seek only the hand of the machinist in the design of the machine. Further, the economy is maintained by a set of mechanisms or forces: the propagation, conservation, and destruction of plants and animals as expressed by the patterns of their distribution and by their interactions with one another.[9]

Stephen A. Forbes, for his part, helped give form to the notion of biotic community that would dominate the development of ecology in the first decades of the twentieth century. Some features of the biotic community echo the Linnaean notion of economy: it is a unit of ecological organization understood in terms of the abundance and distribution of different species and their relations to one another. Forbes's study of lakes also showed a strong tendency toward holism; he urged "the necessity for taking a comprehensive survey of the whole as a condition to a satisfactory understanding of any part."[10]

To the extent that he considered lakes to be like organisms, Forbes approached them with something very much like the spirit of physiology, which included mechanist leanings and an emphasis on detail. He was convinced that the "beneficent order" of the whole was maintained by a "community of interest" among the parts, and he devoted himself to studying the "organic relations" among living things and the manner in which those relations themselves constituted and maintained the balance of life in the lake. To study the black bass, he argued, it was necessary to study all of the species on which it depended and the conditions on which they depended; further, all of the species with which the black bass competed had to be taken into account as well. "By the time [the naturalist] has studied all of these sufficiently he will find that he has run through the whole complicated mechanism of the aquatic life of the locality, both animal and vegetable."[11]

There are clearly important similarities between these two views, but there is a much more important, if less obvious, difference. Linnaeus's conception of the economy of nature was very much a part of the tradition of natural theology that accompanied early mechanist theories, according to which the goal of natural history was to provide evidence for the existence of God. In the eighteenth and nineteenth centuries, however, natural history parted company with natural theology, and God became less and less important in accounts of the natural world. Forbes's study of lakes was a product of this newer, more fully secular, natural history.[12] The transition from one to the other established the trajectory along which ecology would continue to develop through the twentieth century.

Linnaeus adopted John Locke's empiricism as his method for studying the details of the natural world. Following Locke, he assumed that the human mind starts out empty of both fact and theory and attains knowledge only by the impact of sensory data which fall onto the mind like chalk marks on a blank slate. Following this model, Linnaeus imagined a man who falls to Earth, endowed with a full array of sensory and, apparently, intellectual capabilities but with no prior experience of this Earth or its creatures. At first, Linnaeus claimed, this man would perceive only a war of all against all; but as he observed more and more, he would come to perceive an "elementary order," a harmonious and cyclical movement of all things. Before long, the mythical visitor would perceive this order as divine in origin and would praise God for his wisdom and for his divine plan that maintains the balance of the natural world.[13]

A guiding principle of natural theology was that nothing in nature is superfluous or, conversely, that everything in nature serves some purpose. When he was confronted with something that was apparently useless, Linnaeus found that it was nevertheless possible to claim that it served some intermediate purpose. So, for example, a certain insect might offer no direct benefit for humans, but it provides food for the birds that delight human ears. Following a chain of intermediate ends involved a serious and detailed examination of the ways in which organisms interact among themselves. However, nothing important was at stake in these studies, since the result had been determined ahead of time. No matter what he found, Linnaeus was able to interpret it as evidence for the balance of nature and for the existence and beneficence of God. These assumptions were unshakable; Linnaeus's only task as a naturalist was to make the presumed order of the cosmos manifest by showing how wonderfully the natural world serves human interests so that God could receive the praise that is due him. Seen in retrospect, the man who falls to Earth is clearly not a blank slate but comes equipped with conceptual filters through which to strain his experience. It is more as though Linnaeus imagined himself falling to Earth.

At about the same time, Buffon, who was perhaps the most prominent figure in eighteenth-century natural history, was setting the stage for a movement away from theology. In his monumental *Histoire Naturelle,* he raised important new questions which would give shape to more than a century of research in taxonomy, zoology, anatomy, the theory of the generation and reproduction of animals, and the history of the Earth—although he did not himself

take up the idea of the economy of nature. In the spirit
of the Enlightenment, Buffon rejected any account of nature that relied on divine providence. His method had
built into it a kind of anti-dogmatism. He believed that
the study of natural history required the union of two
qualities that seem to be opposed to one another: "the
grand views of an ardent spirit that embraces everything
in a glance, and the little attentions of a laborious instinct that does not attach itself but to a single point."
Theory and observation played off one another, informing
one another, but Buffon urged caution in seeking the
"grand views" since there is always a danger of falling into
dogma. "This is the most delicate and the most important
point in the study of the sciences: to know well how to
distinguish that which is real in a subject from that which
we arbitrarily place there in considering it." In a sense,
nothing is sacred here; even the most cherished assumption must be called into account.[14]

This approach to natural history took hold only gradually. It achieved its first major success in the work of
Charles Darwin. By appealing to a simple mechanical
process of variation and selection, Darwin was able to
offer a convincing account of the diversity of life, and
much else besides, without appealing to the doctrine that
God created each species separately. Darwin had presuppositions of his own, but later naturalists and biologists
would further elaborate and refine his theory, sorting out
"that which is real" from that which "we arbitrarily place
there."

Incidentally, some have identified Darwin as a direct precursor of ecology, largely because of the role of environmental forces in natural selection and the conse-
quent concern with what are now seen as "ecological"
questions in *The Origin of Species*. Nevertheless, there is

good reason to hold that his historical influence on the formation of ecology was only indirect. Darwin was concerned with the influence of the environment on the evolution of species across time (the "diachronic" aspect), whereas early ecologists were concerned with the present state of the interactions of animals among themselves and with their environment (the "synchronic" aspect). Conceptually, some of the problems seem to be the same, but the two sciences have moved in different directions for much of the twentieth century; the idea of an "evolutionary ecology" has emerged only recently.[15]

In the face of the secularization of natural history, interest in the broader patterns of interaction among living things receded to become little more than an undercurrent. Without God as the unifying theme and supreme explanatory principle, the Linnaean economy of nature could persist neither as a theory nor as a coherent field of study. In effect, it fell apart into three separate themes that cropped up here and there throughout the nineteenth century: the interdependence of species, the circulation of elements, and the geographical distribution of species. The emergence of ecology can be dated to the reintegration of these themes into a single discipline within a secular conceptual framework.[16]

At the beginning of the nineteenth century, Alexander von Humboldt took up the theme of the distribution of species when he established the geography of plants as a separate field of research. Humboldt studied what he called "associations," or geographically distinct groupings of plant species, each of which he regarded as a unified whole with its own unique "physiognomy" determined by the proportions and distributions of the various plant species within that region. This marks a significant shift of focus away from efforts to account for the whole of earthly creation toward efforts to understand more local and mundane phenomena. While Humboldt worked to classify associations into types, he was most interested in correlating each type of association with the physical conditions in which it is found in the hope that he could then explain both the character of various associations and the divisions between them. Using various techniques of measurement, including a special instrument called the "cyanometer" that measured the blueness of the sky, Humboldt set about making detailed observations of the environmental conditions that determined the distribution of plants. Most of his *Essai* is taken up with experimental data gathered during his exploration of South America, including a comparison of the vegetation at different altitudes in the Andes and in the Alps. This line of research would be taken up by a number of other geographers throughout the nineteenth century.[17]

Unlike Linnaeus, Humboldt believed that it is nature itself that establishes the order of the world, including the distribution of plants; God plays no explicit role. Even so, his vision of nature shows Humboldt's affinity for Romantic nature philosophy, which seems to have served him as a semi-secular stand-in for divine providence. Like the Romantic biologists of his day, Humboldt did not follow the lead of nature philosophy to the point of rejecting observation and experiment in favor of speculation or aesthetic sensitivity.[18] Even so, because he did not (and perhaps could not) question his assumption that associations are unified wholes, he was limited to describing the surface appearance of those wholes and to drawing mere correlations between their surface appearance and the environmental conditions in which they are found. He can hardly be blamed for this, since he did not yet have the theoretical tools with which to account for the internal structure and functioning of plant associations in terms of the organisms that make them up.

What made the difference for Forbes and his contemporaries was the development of physiology,[19] which played two distinct roles in early ecology. First, a better understanding of the physiology of individual organisms made possible a better understanding of the distribution and abundance of a species. What a living thing needs and what it can do—its physiological capabilities and limits—determine the conditions under which it can live, which goes a long way toward explaining Humboldt's correlations. Second, the methods and concepts of physiology provided a model for reinterpreting the order of nature itself, or at least the internal structure and functioning of larger units of ecological organization, in organic terms. Biotic communities and ecosystems could, at least in principle, be studied as if they were organisms in the physiological, not the speculative, sense. Ecological systems are wholes that maintain their own integrity, but they can be disarticulated in order to understand how they do so.

This integration of natural history with physiology is the beginning, not the end, of the history of ecology. While Forbes and his contemporaries posed a new set of questions about the interactions among organisms, their answers were far from adequate. Physiology would advance significantly in the twentieth century, and more sophisticated models would be needed for the structure and functioning of biotic communities. The idea of the balance of nature itself remained an unquestioned assumption for early ecologists, one that would not be significantly challenged until the closing decades of the twentieth century.

ECOLOGY AND THE ADAPTATION OF METAPHORS

At least where the study of larger units of ecological organization is concerned, the development of ecology since Forbes seems to follow a consistent pattern. A model is proposed whereby units of organization are viewed as being like an organism in some particular way. The model is usually somewhat speculative, and its first presentation is often overstated and dogmatic. Because scientific inquiry is a social effort, criticism arises from all corners, dominated by accusations that the model is "teleological" or "holistic." In response, researchers develop more traditionally mechanistic or reductionist models to explain the same phenomena. The result of this (usually messy) process is some sort of progress. At the very least, the original model is refined and restricted in scope; often, the final form it takes bears little resemblance to the speculative vision from which it sprang. In some cases, the process leads to the development of an entirely new model that covers the same phenomena more adequately.

Frederick Clements, one of the pioneers of ecology in the twentieth century, is best known for his proposal that a plant association, which he called a "formation," is a kind of superorganism. "The formation," he wrote, "is a complex organism, which possesses functions and structure, and passes through a cycle of development similar to that of the plant."[20] One of Clements's great innovations was his recognition that the community is dynamic, perpetually undergoing change or "succession"; he held that this process of development culminated in a well-defined climax state, just as a plant grows up to the mature form appropriate to its species.

Clements appealed not to the organic principle of speculative philosophy but to the organic principle of physiology. In fact, he insisted on the essential identity of ecology and physiology, anticipating their ultimate merging in a single discipline—probably under the name of physiology. Although Clements is often said to have espoused holism, he understood organisms in terms of mechanistic reductionism, and he wanted to account for organic changes in terms of simple stimulus-response reactions. His innovation lay chiefly in his extension of physiological methods, as he understood them, beyond individual living organisms. In so doing, Clements developed new instruments (including the quadrat, which is still used for studying the distribution of plants) and experimental techniques for studying the inner workings of the phenomena of succession.[21]

The belief in the organism concept of community was by no means universal at the time, however, and it might be fairly said that Clements was more influential because of the controversy he touched off than for any of his own direct contributions to knowledge of biotic communities. Henry Alan Gleason was among the strongest early critics of Clements's "superorganism" concept. While he agreed that plant associations exist, he charged that many definitions of associations, especially the organicist view, "extend largely beyond the bounds of experiment and observation and represent merely abstract extrapolations of the ecologist's mind."[22] Associations are nowhere so uniform and so enduring that they can be studied as discrete units and classified into types, Gleason argued; the makeup of vegetation varies constantly in both time and space.

Gleason proposed an "individualistic" model of plant communities, which he regarded as more consistent with observation and experiment. In this model, the soil of any given area will contain a wide variety of seeds, some of which can thrive, or at least survive, under current conditions, some of which cannot; as environmental conditions vary in space and time, so will the makeup of the association. "The vegetation of an area is merely the resultant of two factors, the fluctuating and fortuitous immigration of plants and an equally fluctuating and variable environment."[23] This may be as much an exaggeration as Clements's insistence on the unity of plant formations, and it may be founded in a naive conception of the relation of theory to observation and experiment, but Gleason's criticism proved to be useful in that it helped to bring Clements's own exaggerations under more careful scrutiny.

Writing some years later, Arthur Tansley echoed Gleason's criticism of the superorganism concept. He began his pivotal 1935 article "The Use and Abuse of Vegetational Concepts and Terms" with a broad critique of the tendency of Clements's supporters to fall back into the "expression of a creed—of a closed system of religious or philosophical dogma." Tansley did not object to the organism metaphor as such so long as it was carefully qualified; he suggested the term "quasi-organism" as both more cautious and more useful: to say that a community *is* an organism is to express faith in holism rather than to state a scientific theory. For Tansley, the problem with Clements came down to a matter of the meaning of the organic principle in the context of ecology.[24]

Tansley's response to Clements was more than just semantic, however. Tansley was primarily concerned that the concept of community itself was too narrow to account for all observations about the relationships among living things. He proposed what he took to be a more adequate concept, the ecosystem, which he defined as "the whole system (in the sense of physics) including not only the organism-complex, but also the whole complex of physical factors forming what we call the environment of the biome"; such systems were "the basic units of nature on the face of the Earth."[25] Although he coined the term, Tansley did not develop the ecosystem concept further nor did he undertake the scientific study of ecosystems. Nevertheless, the publication of his paper was one of the watershed events in twentieth-century ecology.

The final result of the controversy over Clements's community concept, then, was significant scientific progress in ecology, in terms of the generation and refinement of useful concepts and theories. First, the organism metaphor was explicitly reduced to the role of a heuristic device that could be useful for conceptualizing complex phenomena and for stimulating research but that was not under any circumstances to be taken too seriously. Even Clements was more cautious in his use of the organism metaphor in his later work, though he never abandoned it altogether.[26] Second, and most important, this episode led to the development of a new and important ecological concept.

When Raymond Lindeman took up and developed the ecosystem concept in the 1940s, his efforts opened up a new way of interpreting the organic principle. For Clements, communities were like organisms because they develop (by a process of succession) toward a stable, specific form. For Lindeman, on the other hand, ecosystems were like organisms because they have their own internal process of metabolism: energy flows through ecosystems and is either exploited or wasted, just as energy is assimilated or wasted by individual plants and animals. While the community approach studied only the abundance and distribution of species, the new "trophic-dynamic" approach looked beyond organisms themselves to "the relationship of trophic or 'energy-availing' relationships

within the community-unit to the process of succession."[27]
This opened myriad new possibilities for ecological research that incorporated new tools and techniques. Even
so, the development of this use of the ecosystem concept
bears a striking resemblance to that of the community
concept.

The 1960s and 1970s saw the emergence of what its
adherents called "the new ecology," centered on the work
of two brothers, Eugene and Howard Odum. They turned
to new models and techniques of systems analysis, including cybernetics, information theory, game theory, and
computer simulation. Following Tansley's claim that the
ecosystem is the basic unit of life on earth, the new ecologists unilaterally redefined ecology as the study of the
structure and functioning of ecosystems.[28] Despite the
persistence of the community and population approaches,
proponents of the new ecology came to believe that the
systems approach would serve as a kind of unified field
theory of ecology, since systems theory could be applied
to natural order at any scale. In the third edition of his
classic textbook *Fundamentals of Ecology,* Eugene Odum
presented a hierarchy of natural systems, from genetic
systems to ecosystems, all of which were equally "regu-
larly interacting and interdependent components forming
a unified whole."[29]

The confidence of the new ecologists in the power
of systems theory, and the methods by which they carried out their research, was based on their firm belief in
the holistic perspective: ecosystems were to be studied
from the top down, from the whole to the parts. One expression of this was the belief that ecosystems are best
studied through what Howard Odum called the "macro-
scope," which would eliminate details in order to gain
knowledge of the system as a whole. In this view, the

components of systems could be treated as black boxes whose only significant features were chemical and energetic inputs and outputs. Further, interpreted as cybernetic systems, the parts of the system could be seen as serving functional roles in the maintenance of that whole in a manner directly analogous to the organs of an animal body; an organism and an ecosystem were regarded as having equivalent degrees of functional integration, expressed as "self-maintenance" and "self-regulation." Eugene Odum went so far as to hint at goal-directedness in ecosystems. In his 1969 paper "The Strategy of Ecosystem Development," he wrote of the development or succession of ecosystems as "reasonably directional," which implied that the biotic community as a whole used "strategies" to modify its environment.[30]

Critics pointed out that the strength of the holistic systems approach was also its greatest weakness; namely, its tendency to underestimate the importance of details in the study of ecological interactions. As Hagen wrote, "Treating trophic levels as black boxes with energy inputs and outputs had proved a powerful investigative tool, but when pushed too far it could also mislead. . . . The living world simply did not operate that way."[31] The attempt to "reduce" the complexity of biological systems to simple thermodynamic "wholes" left out much that was important for any understanding of how the living world operates. As one critic put it, "pesky biological details matter a lot."[32] As was the case with Clements, the new ecology seemed to have gone astray because it had adopted holism as a matter of faith, or at least because it stated its objectives with the "flavor" of a creed or dogma.[33]

By the end of the 1960s, it was becoming clear that ecology would not, and perhaps could not, rally around the systems approach alone, nor could it simply regard ecosystems as the basic unit of nature. In truth, the discipline was becoming increasingly divided. The consequences of this division are most clearly seen in what Robert McIntosh calls the period of "introspection" that has dominated ecology since the mid-1970s, as if the science were groping for a renewed sense of its identity after the systems approach took its place as just one among many approaches to ecological research.[34]

There have been two kinds of response to the "identity crisis" in ecology. On the one hand, there are calls for ecology to become a more rigorous science according to this or that model of what science ought to be.[35] On the other hand, there are calls for pluralism in ecology that emphasize the usefulness of a number of different— and non-overlapping—approaches to the study of the natural world. One may study the physiology of a particular species in light of its interactions with its environment (physiological ecology), the interactions of predator and prey (population ecology), the abundance and distribution of species as they interact within a given area (community ecology), or the cycles of energy and matter through physical systems of which organisms are a part (ecosystem ecology). Results from one approach to interactions at one scale cannot necessarily be carried over to another approach or even to a similar approach at a different scale.[36]

One further consequence of the breakup of new ecology has been the emergence of what are called "non-equilibrium" approaches in ecology. These are efforts to model ecological interactions without assuming that there is an idealized steady state or integrated whole to which those interactions contribute. In effect, the idea of the balance of nature is coming more sharply into question than was ever possible before. In *Discordant Harmonies*, for example,

Daniel Botkin examines prevalent metaphors for nature—
nature as divine order, nature as organism, and nature as
mechanism—and rejects them in favor of an *informational*
metaphor. One characteristic of this approach is its acceptance of the randomness of natural processes: "At the level
at which organisms respond to and affect their environment, the world is one of risk, predictable only to proba-
bilities. Nature as perceived by living things is a nature of
chance."[37] This last is a sweeping claim on Botkin's part,
and it is not entirely clear he is justified in making it ex-
cept, perhaps, on rhetorical grounds. Nevertheless, he insists that good management requires recognition that ecological systems are characterized as much by disruption
and change as by stability.

This is the first strike against efforts to turn to ecology
to bolster the claims of speculative environmentalism. The
organic metaphor might well persist in ecology, but it
bears little resemblance to its speculative ancestor and it
does not hold absolute sway. Properly speaking, ecological
theories are neither organicist nor mechanistic; they are
increasingly sophisticated mixtures of those and other
metaphors that possess some degree of predictive power.
Further, the scope of the theories is considerably nar-
rowed; they are useful in their context, but it is not at all
clear that they have relevance in other contexts. Indeed,
many of the models used by ecologists cannot usefully be
borrowed by other ecologists studying a different level of
living interaction, let alone by philosophers seeking to inform an ecological worldview.

SELECTIVE SCIENCE

Even if the theories and concepts employed by ecologists
were unambiguously organicist, it might not be such a
good idea for environmental philosophers to appropriate

those theories and concepts for their own purposes. At the very least, they must first answer two sets of questions about the scope and limits of scientific knowledge. The first set touches on a long-standing debate about the status of theoretical entities: Even if a scientific theory is successful, what does it tell us about what is really going on in the world? Can scientific inquiry produce metaphysical truth? The second set touches on a related debate over what has been called the problem of demarcation: What is included in the domain of the natural sciences and what is not? What is the proper relationship of the natural sciences to other domains of human thought and practice?

In one common account, scientific laws describe regularities in natural phenomena, while theories explain those regularities in terms of entities and forces that cannot be observed. Boyle's law, for example, is a general description of the observed relationship among the volume, pressure, and temperature of gases. This regularity is explained in terms of the kinetic theory of gases, according to which a gas is made up of atoms moving about with greater or lesser energy. In the traditional debate over the status of theoretical entities, realists contend that the kinetic theory of gases must have tapped into underlying reality: atoms really do exist, and their movement really does account for the behavior of gases. This they see as the only way to account for the theory's success. Instrumentalists, on the other hand, contend that it does not matter whether the theoretical entities really exist or not, so long as the theory is useful for making further predictions about experience; the success of the theory implies nothing whatsoever about the underlying reality.

Environmental philosophers could use the sciences to make headway toward an ecological worldview only if a fully organicist account of ecosystems were both success-

ful and given a realist interpretation. If the new ecology
had been as successful as the kinetic theory of gases, and
if scientific theories were known to provide reliable insight into reality, then it might be claimed that there really
are ecosystems and that they really are like organisms in
some important respects. From that starting point, environmental philosophers could then go on with some confidence to draw out the ethical and political consequences
of human participation in (or disruption of) ecosystems.

Ecology, as my account has shown, cannot even begin
to meet all of these conditions. Aside from the fact that
ecological theory is not organicist and that it has not met
with the same kind of success as its physical and chemical
counterparts, there are reasons to doubt the realist interpretation of scientific theory. To put this the other way
around, there is no more reason to believe that atoms are
really real than there is to trust in Hegel's speculation, and
the same applies to ecosystems. This is not to say that the
kinetic theory of gases is not a well-established and useful
scientific principle or that ecosystem theory may not turn
out to be such. It is only to say that, from a skeptical point
of view, the question of the ultimate reality of atoms and
ecosystems may well be irrelevant: it does not add anything to, or take anything from, the usefulness of either
theory.

In general terms, the sciences can answer some questions about the natural world, but only tentatively and
only in the context of continuing controversy where even
the facts themselves are in question. The sciences offer
prediction and control: they make it possible to see what
is likely to happen under given circumstances, and they
make it possible to construct powerful tools for getting
things done. Within these limits, there may be important
lessons to learn from the sciences—especially about the

possibilities, limits, and consequences of human actions in various circumstances. However useful this might be, the tentative truths of the sciences rest within a much larger and deeper uncertainty. However good we may be at predicting what happens next, the "secret springs and principles" of the universe, and our proper place within it, will remain hidden.

The problem of demarcation casts a different light on the relation of philosophy to speculation. Hume used his skeptical principles, informed by empiricist assumptions, to distinguish knowledge from nonsense. If a book contains neither "abstract reasoning concerning quantity or number" nor "experimental reasoning concerning matter of fact and existence," he argued, then it should be committed to the flames, "for it can contain nothing but sophistry and illusion."[38] The logical positivists of the early twentieth century updated Hume's skeptical empiricism by integrating it with modern propositional logic. In the process, they introduced the contemporary version of the problem of demarcation.

In either case, the line of demarcation is drawn to follow the contours of an especially narrow conception of what can count as knowledge. For logical positivists, statements have meaning only if they are verifiable, and a statement is verifiable if and only if there is some observation statement that is relevant to its truth or falsehood. The upshot of all this is the rejection of all metaphysics: claims about ultimate reality, which can be neither confirmed nor denied by experience, are simply meaningless. Like Hume, positivists assumed that the secret springs and principles of nature are hidden from scientific inquiry, and because scientific inquiry is the only valid form of inquiry into the nature of nature, there are no other means for finding them. In spite of this crucial limitation, positivists could at least claim that the sciences can answer every question worth asking—since the only questions worth asking are those that can be answered scientifically.

While it makes perfect sense to distinguish between the sciences and other domains, I disagree with the positivists as to the nature and location of the line of demarcation. The positivist conceptions of language and of human consciousness are simplistic, and as a result the criterion of meaningfulness is far too restrictive. A range of non-scientific statements can also be meaningful, even if their truth or falsehood cannot be established by rigidly empirical means. Further, as I have already shown in the case of ecology, the boundary between science and other domains is permeable. Ideas and images flow back and forth between the sciences and the array of metaphysical, religious, ethical, social, aesthetic, and personal considerations that make up the rich texture of human life. It will do little good, then, to follow the positivists in elevating the sciences and dismissing all else as nonsense. Instead, the challenge is to give each domain its due while examining the interactions among them all.

I am interested in one particular kind of interaction. The natural sciences now wear the mantle of authority in all matters concerning knowledge of the material world. Environmental philosophers often borrow elements of scientific theory, apparently hoping that some of the authority of the natural sciences will rub off on their metaphysical and ethical claims. This practice raises a number of questions. To what extent is such borrowing legitimate? What happens to a scientific theory when it is translated to another domain of human thought and practice? Does such translation preserve that which gave the theory its aura of authority in the first place?

Many environmental philosophers relate to the natural sciences as they relate to the philosophical tradition: they pick and choose, seeking out ideas that seem to bolster their own claims on behalf of relatedness. Selective science, like selective philosophy, has its dangers. Ecological concepts and theories have been refined and delimited by more than a century of scientific inquiry. While metaphysical organicism is universal in scope, whatever elements of organicism remain in ecology are not only tempered by their mixture with other metaphors but are also limited to a heuristic role in conceptualizing particular terrestrial ecosystems. The power and authority of an ecological theory is proportional to the degree to which it has been tempered and limited in this way. Once it has been refined to fit a particular context, the organic metaphor cannot be lifted back out again and carried to other domains—or merged with metaphysical organicism—without losing much of its specificity and, as a consequence, all of its authority. That said, let me examine the practice of selective science in some detail.

Like selective philosophy, selective science reveals the essentially reactive character of much environmental philosophy. Practitioners cast about for anything that might be useful in their opposition to the dominant paradigm, weaving diverse and sometimes contradictory strands into their ecological worldviews. Anything might suggest a potential alliance; sometimes the mere presence of certain key words is enough. When Frank Golley, an ecologist, considers the compatibility of the basic principles of deep ecology with ecology, he concludes that

> the two ultimate norms or intuitions of deep ecology coincide with ecological understanding. The language in which these premises are presented is misleading. 'Self' and 'equality' are employed in unconventional ways but with interpretation

the meaning of these words can be understood in an ecological context. Thus, I conclude that these premises of deep ecology do not conflict with the observations and conclusions of the field ecologist.

Don Marietta concurs, noting that "certain words used in ecological descriptions are the same words used in speaking of environmental values . . . such words as *stability, diversity, unity, balance, integrity, order,* and *health.*" This does not strike him as a mere coincidence; rather, "the different uses of the same terms reflect the fusion of fact and value." Such a fusion can only take place within the context of a particular worldview, but he insists that, in that context, "our obligation toward the environment can be grounded in ecological principles, and this grounding is as sound as that available to any other ethical approach."[39]

For others, the connection between science and speculation is far more substantial—but no less selective. When he considers the possibility of using concepts from quantum physics for the purpose of bringing about a new worldview, J. Baird Callicott makes a revealing statement:

> There are a number of respectable metaphysical interpretations of quantum physics from which an environmental philosopher might choose. The venerable and conservative Copenhagen Interpretation portrays ultimate physical reality, as we experience it indirectly, to be a marriage between subject and object.

Certain versions of ecology and quantum physics, he maintains, can help to break down the old, mechanistic worldview which separated object from object and subject from object; environmental philosophers need only select one that will best suit their own purposes. He does not know what future historians will call the paradigm that he hopes is emerging from the breakdown of the subject/

object divide, or even what form it might eventually take. He is only too willing to provide names for it, however: "the organic worldview, the ecological worldview, the systems worldview."[40]

Callicott's appropriation of the sciences is not just selective but actually preemptive. He has placed all his bets on the paradigm he hopes will emerge from future scientific research. A paradigm shift, according to Thomas Kuhn, is a revolution in science in which even basic presuppositions, standards of evidence, and so on, are thrown out and replaced by the relevant community of researchers. By this definition, paradigm shifts are not wholly rational, because the standards of scientific rationality themselves are up for grabs. Then again, if Kuhn is correct, the direction of a scientific revolution is not to be dictated by those outside the scientific community: such revolutions happen from within, based on the demands and contingencies of research rather than on the demands of environmental politics. In any case, neither the occurrence nor the outcome of a scientific revolution can be predicted in advance.

Some environmental philosophers seem to rely on selective science as a way to overcome their own profound ambivalence toward the sciences. Bill Devall and George Sessions, who explicitly cite ecology and physics as sources of inspiration, also insist that the basic principles of deep ecology "cannot be validated, of course, by the methodology of modern science based on its usual mechanistic assumptions and its very narrow definition of data."[41] At this point, the radical and speculative projects of environmental philosophy come directly into conflict. The speculative project seems to require some appeal to the sciences to inform and justify claims to knowledge about nature, but the radical project casts suspicion on anything that comes from a method of inquiry that is supposedly based on the very mechanistic assumptions that are complicit in environmental degradation. This places Devall and Sessions in an odd position: as a science, ecology seems to fall on both sides of the ideological fence, and they must somehow make an ally of what they themselves see as their most dangerous foe.

Arne Naess strikes the balance by insisting that while ecology has "suggested, inspired and fortified" the deep ecology movement, the movement is really "ecophilosophical" rather than ecological. Ecology, he notes, is a science limited to the use of scientific methods, while philosophy "is the most general forum for debate on fundamentals, descriptive as well as prescriptive." Again, the connection between the two often seems to be a matter of semantics: deep ecology adopts the language of ecology to inform its relational worldview. On the side of ecologists, Frank Golley admits that deep ecology cannot be *derived* from ecology, but he maintains nonetheless that there are "bridges" or "points of compatibility" between the two such that "deep ecology norms can be interpreted through scientific ecology." He finds this compatibility in the use of terms such as "ecological hierarchical organization" and "the exchange of energy, material and information." What he seems to mean by this is simply that despite semantic problems and a few qualms about the sweeping claims of deep ecologists, deep ecology is not explicitly contradicted by the scientific study of the living world.[42]

The phenomenological approach once again offers an experiential bypass of the problem and demonstrates how one can embrace and reject the sciences at a single stroke. David Abram, for example, seems to hold out the possibility that scientific knowledge can be achieved, or at least replicated, by non-scientific means. A perceptual immersion in the surrounding world, he maintains, leads to the discovery of the *biosphere:*

> the matrix of earthly life in which we ourselves are embedded. Yet this is not the biosphere as it is conceived by an abstract and objectifying science, not that complex assemblage of planetary mechanisms presumably being mapped and measured by our remote-sensing satellites; it is, rather, the biosphere as it is experienced and *lived from within* by the intelligent body.

Hence, Abram sees in phenomenology the possibility of an "ecology from within," an encompassing relatedness or reciprocity between the self and the biosphere.[43]

Later, it becomes clear that this "ecology" is based on an analogy: language, the world of primary perception, and the biosphere are all "complex webs of relationships," which Abram seems to collapse together into a single relatedness. Specifically,

> The notion of earthly nature as a densely interconnected organic network—a 'biospheric web' wherein each entity draws its specific character from its relations, direct and indirect, to all the others—has today become a commonplace, and it converges neatly with Merleau-Ponty's late description of sensuous reality, "the Flesh," as an intertwined, and actively intertwining, lattice of mutually dependent phenomena, both sensorial and sentient, of which our own sensing bodies are a part. It is this dynamic, interconnected reality that provokes and sustains all our speaking, lending something of its structure to all our various languages.

Because the sensuous world itself provides the "deep structure of language," and because the sensuous world "converges neatly" with the "commonplace," that is, the scientific, understanding of the biosphere, Abram is able to imply that, in order to understand our earthly environment, we must immerse ourselves in primary perception and re-awaken the evocative and relational power of speech.[44]

In effect, he presupposes the truth of what has been discovered only with the help of the "remote-sensing satellites" he disdains: the biosphere as a complex web of (physical and chemical) relationships. He calls it a "commonplace," glossing over the long tradition of scientific inquiry and technological development that makes the modern notion of the biosphere possible. This echoes the problem that Darwin would have posed for Hegel, had the latter lived. Recall that one option for Hegel would have been to update his nature philosophy to fit the new theory even as he dismissed the mode of inquiry that made the theory possible. It seems as though Abram is simply being disingenuous, borrowing a concept that can only have been the product of centuries of scientific inquiry while pretending that it can be discovered independently by direct perception.

The shortcomings of selective science are not limited to the appropriation of ecological concepts and theories. They apply also to the efforts of Warwick Fox and others, especially the tellers of the universe story, to draw from evolutionary biology and modern cosmology to bolster their belief that everything that exists is really part of a single unfolding order. In a sense, such thinkers have recast the balance of nature in dynamic terms, harking back to an older use of the term "evolution." Before Darwin, evolution referred to the progressive unfolding of a pre-existing internal organizing principle, especially as exhibited in embryonic development. The process is a passage from a state of simplicity to one of complexity. Fox and others pick up on the purposive and directional implications of this older understanding to imply that the development of the universe is guided by a kind of creative energy, the intentions of which are to be respected.

Darwin's theories, however, represent a radical revision of the meaning of evolution. Their power is precisely in the central tenet that it is a mechanism, natural selection, which drives the proliferation of life on earth. There is a wide range of variation within any group of organisms, and some of those variations will give a slight reproductive advantage to the individuals that possess them relative to the prevailing environmental conditions. Over time, as the successful variations become more common, the average character of the population will change. If two sub-groups of the population are exposed to different environmental conditions, they will diverge, a process that leads ultimately to the formation of new species. While it may be the case that individual organisms are engaging in goal-directed activity as they struggle for survival and reproductive success, the process of evolution is essentially directionless.

There are those who will protest that evolution produces beings of greater and greater complexity, so that complexity itself can be taken to be the central tendency of evolution. It is ironic that some environmental thinkers would draw from this kind of purposive reading of evolution to bolster their ecological worldviews. If strong forms of anthropocentrism, those that insist that humans occupy a place of central importance in the cosmos, are to be rejected as inadequate for environmentalist purposes, then the belief that evolution tends to produce complex beings should also be rejected. If evolution tends toward complexity, and if it is observed that humans are among the most complex of organisms, then it is not a very great leap to conclude that humans are the inevitable and high-

est product of evolution. Indeed, in the first flurry of reaction to Darwin's work, many people settled on this kind
of argument as a safe and comfortable way of absorbing
the impact of natural selection. More recently, tellers of
the universe story have identified humans as the emerging consciousness of the unfolding cosmos. "The human
is that being in whom the universe comes to itself in
a special mode of conscious reflection," writes Thomas
Berry, noting that many scientists now grant "that some
form of intelligent reflection on itself was implicit in the
universe from the very beginning."[45] This is a rhetorical
move that echoes Hegelian nature philosophy more than
it does Darwinian natural selection.

There are a number of reasons to doubt that evolution tends toward complexity. The vast majority of living
things are actually very simple. Statistically speaking, the
mode of life is bacterial. If there is a tendency, it is toward diversity, not complexity. In a recent essay, Stephen
Jay Gould offers a statistical analysis that leads him to
conclude that some complexity is a likely side effect of di-
versification but that no particular form of complexity is
inevitable. If the geological clock could be set back to the
first emergence of life on Earth and allowed to run forward again, Gould argues, the results would probably be
very different. This undermines the comforting image of
a smoothly and rationally unfolding cosmos, just as it undermines the comforting notion that humans stand at the
pinnacle of evolution; there is no such pinnacle.[46]

The limits of scientific inquiry, combined with the
limits of speculative nature philosophy, cast doubt on the
first claim of speculative environmentalism: environmental philosophers cannot assert with confidence that the
natural world is fundamentally relational. More than this,
it seems as though the fundamental nature of nature may

remain hidden from human understanding regardless of the method used. As it happens, ecology and evolutionary biology may yet have profound implications for the self-understanding and the practices of human beings, but not in the way environmental thinkers have hoped.

It is important to remember that the goal of environmental philosophy is to change the way people behave, preferably by means of some sort of ethical injunction. In a sense, the second claim of the central case is more important than the first: humans have a moral obligation to respect and preserve the order of nature. If the doubts I have raised are decisive, if the organicist or relational claims of environmental thinkers are to be rejected, then this second claim is left without what was intended to be its foundation. Nevertheless, the idea that we may have some sort of obligation to the natural order deserves careful consideration on its own merits.

Part 2

Obligation

[3]

A Place on Earth

In 1765, Jean-Jacques Rousseau was driven into exile following the publication of *Émile*. During that first summer of his wanderings, he lived for a short while on the island of Saint-Pierre in the Lake of Bienne. One of his favorite pastimes while there was to row out to a smaller uninhabited island in the lake where he could be alone. On one of these visits, he decided that this small island would be "a very suitable home for rabbits, which could multiply there in peace, without fearing or harming anything." At Rousseau's insistence the steward procured some rabbits, and the human inhabitants of the island of Saint-Pierre proceeded "in great ceremony" to introduce them to the smaller island. Rousseau was pleased to note that the rabbits had already begun to breed before he left for the mainland, and he was sure that they would flourish. He rhapsodized: "The founding of this little colony was a festival day. The pilot of the Argonauts was not prouder than I, when I led the company and the rabbits in triumph from the large island to the small one."[1]

What is most striking about this story is that Rousseau loved nature and wished for nothing more fervently than to live under nature's tutelage. This love was not, however, enough to prevent him from carrying out a small feat of ecological engineering that may well have resulted in a minor biogeographical catastrophe—at least from the rabbits' point of view.

There may be no way to know what actually happened to Rousseau's rabbits. I have not found any written record of their fate and, ironically, the smaller island has all but disappeared since Rousseau's time, its soil having been carted off to repair storm damage to the larger island. However, the conventional wisdom of game management and island biogeography suggests that a population of a prey species living without predators tends to be very unstable, subject to the now familiar pattern of overshoot and crash. It is entirely possible that the rabbits thrived for a time, their numbers growing, until they overwhelmed the carrying capacity of their small habitat. At that point, having devastated the lovely vegetation Rousseau admired so much, the rabbits would have starved to death.

In hindsight, Rousseau's triumphant procession was clearly imprudent; had he known what we know, he might have chosen differently. Knowledge alone is not the issue, though, and neither is prudence. Looking back on Rousseau's actions, it is easy enough to claim that our knowledge of the natural world is superior to that of the eighteenth century, at least in that it allows us to predict more of the consequences of certain kinds of action. What remains to be seen is whether we can claim to have a better standard of conduct.

Rousseau's apparent goal in establishing his rabbit colony was to make the smaller island more pleasant. He clearly enjoyed the thought of rabbits living in peace, without fear, much as he would have lived had he been given the chance. Perhaps he also enjoyed the thought of all that vegetation being put to good use. In other words, he was acting on the basis of his own understanding of what would be best for a place he loved, a place he would remember fondly until the end of his life. At best, he was heeding a noble (if misdirected) aesthetic impulse; at worst, he was simply being self-indulgent.

In either case, it could be argued that Rousseau suffered a kind of moral blindness. He failed to consider the consequences of his actions for the rabbits and for the plant community of the island, not only because he was ignorant of ecology but also because their well-being simply did not matter to him. Rousseau claimed to be a student of nature, and he insisted time and again that, if nature is good, then so must be the natural impulses of human emotion. As a consequence, he did as he wished and went on his way with a clear conscience. This may be an uncharitable reading of Rousseau, but it could be argued nonetheless that, had he been guided by a better moral standard, he might have chosen simply to enjoy the island and leave it as he found it.

Finding a better moral standard is precisely the problem for environmental philosophers who adopt the speculative path. If there is to be a standard to govern the relationship of humans to nature, what is it, and where is it to be found? More pressingly, if there is disagreement over moral standards, to what might we appeal to bring about an agreement? Suppose that the steward of the island of Saint-Pierre had disapproved of Rousseau's scheme on the basis of some competing standard of value. Is there something about nature or about the human relationship to nature or about the place itself that could have decided the issue?

For many environmental philosophers, the answer to this last question is "yes," or at least "we hope so." The strongest version of the argument runs as follows. If nature is analogous to an organism in that it has its own internal principle of organization, and hence its own goals and interests, then it might be possible to think of Homo sapiens as analogous to one of the organs with its own specific function within the whole. If a liver or a kidney does not contribute its share to the overall well-being of the organism, it is regarded as diseased or simply bad. If the analogy holds, then particular humans or groups of humans may be judged to be good or bad based on whether they are functioning well or poorly within the systems of which they are a part. Very few environmental thinkers would make the claim in this strong a form, but the effort to redefine humanity as part of something larger than itself is a common theme in environmental thought.

Aldo Leopold, for example, appealed to the notion of community to establish the connection between humans and nature. He wrote that "a land ethic changes the role of *Homo sapiens* from conqueror of the land-community to plain member and citizen of it. It implies respect for his fellow-members, and also respect for the community as such." In this case, the community in question is the biotic community, an association of plants and animals in a given area, in which the abundance and distribution of each species is regulated by the abundance and distribution of all the other species. The community can also be understood as an ecosystem by following the flows of energy and matter through the land pyramid.

It is important at this point to recall that the concept of biotic community is based on a metaphor from human communities: just as humans play roles within their various communities, so organisms play roles within the natural systems of which they are a part. Both senses of community have been informed by the organism metaphor, to the effect that communities as a whole are thought to have interests that transcend the interests of any individual. Leopold used this complex interplay of metaphors to gain moral ground: humans are "plain members and citizens" of biotic communities, with all of the responsibilities that go with citizenship. This led him to his much-quoted dictum that "a thing is right when it tends to preserve the integrity, stability, and beauty of the biotic community. It is wrong when it tends otherwise."[2] By this standard, of course, Rousseau was clearly in the wrong.

But, the skeptic is bound to ask, is this the correct standard? The answer hinges on a further question: How and to what extent can we humans know our place in the natural order? With this second question, the focus of skeptical environmentalism shifts slightly: while I have been considering whether we humans can attain sure knowledge of nature, I must now consider whether we can attain sure knowledge of human nature. When Leopold cast humans as "plain members and citizens" of biotic communities, he was drawing on a particular vision of what it means to be a human being: we are animals with a given set of capabilities and limitations, with a distinct but limited role to play in ecological systems. More recent environmental thinkers have followed suit; they are convinced, it would seem, that we must know who we are and where we are before we can decide what we ought to do. If their characterizations of humanity should turn out to be unfounded, then their ethical conclusions are cast into doubt—compounding the doubts I have already raised.

When brought under skeptical scrutiny, I will argue, human nature reveals itself to be profoundly ambiguous, far too much so for us to be pinned down to a single, determinate place in the natural world. The consequences of this ambiguity are profound and far-reaching: not only does it cast further doubt on the work of contemporary environmental philosophers, but it also begins to suggest a very different way of understanding environmental problems and the possibility of addressing them.

FREEDOM AND NATURE

Anyone who would argue that humans are part of nature in some morally significant sense must at some point take up the problem of human freedom. Immanuel Kant, for example, maintained that ethics can only exist as a separate field of inquiry if humans have free will. If it were not possible for humans to transcend natural determination, then the study of human behavior would simply be a matter of anthropology which, as a science of nature, could perhaps be reduced to the terms of physics. If this were the case, then it would make no more sense to hold people morally responsible for their actions than it would to hold a volcano morally responsible for erupting. Freedom became a problem for Kant because humans are undeniably natural beings subject to deterministic natural law. How, then, is it possible to conceive of humans as possessing free will?

Kant solved the problem, at least to his own satisfaction, by demonstrating that there need be no contradiction between determinism and free will, as long as it is recognized that nature is "real" and freedom is "ideal." Each is addressed by a different intellectual faculty. I know the world to be determined because natural laws are constitutive principles of the understanding, but the natural world is nothing more than the realm of possible experience as it is shaped by the understanding. Reason, for its part, asserts the principle of freedom, but only as a regulative ideal: it is not that I know that I am free but that I am able to think of myself as if I were free. Because the mind has no access to things in themselves, according to Kant, it is neither possible nor necessary to account for the relationship between freedom and nature; all he needed to establish was that there is no contradiction in asserting both.

This seems straightforward enough, but I suspect that matters are not nearly as simple as Kant supposed. Writing some decades before Kant, Rousseau was really the first to attempt a systematic treatment of human moral freedom in its relation to nature. Whatever his opinion of rabbit colonies, he hoped to harmonize freedom with nature's laws in order to reform what he saw as a corrupt civilization. The difficulties he encountered along the way and, I would argue, the ultimate failure of his project reveal the full complexity of the problem of freedom.

If there is anything that unifies Rousseau's thought, it is his obsession with finding the conditions under which human beings could be happy. He settled on a peculiar kind of freedom which, he insisted, consisted not in doing what one wants but in never being compelled to do what one does not want to do. In short, he argued humans could be happy only if they avoided any dependence on others or any subjection to their whims. The only guarantee of independence was contentment, which Rousseau described as a proportion between desire and power: if I do not want more than I am able to get for myself, he reasoned, then I will never be discontented and I will have no need to be dependent on others or to manipulate them in pursuit of my own selfish interests. Rousseau seemed to believe that it is easiest to maintain this kind of contentment in the state of nature, in which humans are ignorant, innocent, and alone. Lacking even the imagination to desire more than they can get, natural humans lived tranquil lives under nature's tutelage. Nothing could corrupt their natural goodness so long as the natural passions of self-preservation and pity served as their only guide to conduct. As far as Rousseau was concerned, nature intended humans to live in this state of innocence, and it took pains to delay their further development as long as possible.

Rousseau faced a dilemma, however. While he continually looked backward to the lost age of innocence, he also insisted that moral freedom was the essential trait of humanity. It was only with the development of reason that human actions took on moral significance, since it then became possible to act on principle rather than merely on sentiment or instinct. From this perspective, natural liberty was not true freedom at all, since "to be ruled by appetite alone is slavery, but to obey a law that one prescribes to oneself is freedom."[3] Rousseau's dilemma arose because of the danger inherent in moral freedom: with the advent of reason comes the end of contentment. This, for Rousseau, was the fall from grace that brought inequality, vice, and misery into the world and that set up the tyranny of humans over nature. In the extraordinary Note IX in the *Discourse on Inequality*, Rousseau encouraged those who remain uncorrupted by civilization to return to their natural state:

O you to whom the celestial voice has never made itself heard, and who recognize no other destination for your species than to live out this short life in peace; you who can leave behind in the cities your fatal acquisitions, your restless spirits, your corrupted hearts and your boundless desires, regain, if you can, your former and first innocence; go into the woods and lose the sight and the memory of the crimes of your contemporaries, and never fear that you are bringing dishonor to your species by renouncing its enlightenment in order to renounce its vices.

As for himself, Rousseau had to admit that he was no longer capable of such an absolute renunciation: he could "no longer be nourished on grass and acorns, nor do without laws and chiefs."[4]

In his mature political work, Rousseau attempted to resolve this dilemma by finding a way for humans to live together *as humans*—that is, as free moral beings—while avoiding the corruption of civilized life. Natural liberty, the independence that characterizes life in the state of nature, exists only through dependence upon and obedience to natural law. As Rousseau construed it, the social contract was set up so that the members of the body politic could enjoy civil liberty, which also consisted of independence from the arbitrary whims of others. Civil liberty came about when citizens subjected themselves entirely to the general will which is, in effect, human moral freedom in the abstract, stripped of all the particularities and selfishness of any individual will. Because it is impersonal and impartial, the general will represents the genuine self-interest of the sovereign body politic. Because each citizen is a constitutive member of the body politic, Rousseau argued, to obey the general will is, in effect, to obey oneself.

This was the reasoning behind the education of Émile,

which had the aim of preventing the development of any passions or desires contrary to natural human goodness while at the same time preparing Émile for life in human society. At the end of his education, Émile made the paradoxical declaration that his teacher had made him free by teaching him to yield to necessity.[5] According to Rousseau, civil liberty was analogous to natural liberty because it established the conditions for independence and contentment through subjection to a higher law and because it served to prevent the tyranny of humans over each other and over nature.

Even at this stage, when he was at his most optimistic, Rousseau remained suspicious of civil society. There is a persistent sense in *The Social Contract* and *Émile* that, for all its supposed advantages, civil liberty is really a second best when compared with the blessed ignorance of life in the state of nature. Rousseau regarded as inevitable the corruption of even the best of states, because individual wills fight against the general will, seeking always to manipulate others to their own selfish ends.

What this shows is that it is very difficult to find a common ground on which freedom and nature can meet. The truce Rousseau arranged was fragile, and it remained a constant temptation for him to simply turn away from moral freedom, to immerse himself once more in something like the state of nature. Indeed, at the end of his life, during his long exile, Rousseau's cautious optimism collapsed into despair. He renounced, as much as he could, the life of the mind and sought a return to the state of nature as the only option for securing the little happiness he could find. His happiest moments came to him during the two months he lived on the island in the Lake of Bienne, during which time he was pleased to avoid any of the intellectual activity that had brought him such misery, and he was still more pleased to pass his days without any sort of plan or desire. He drifted from one activity to another, occasionally losing himself entirely in a state of timeless bliss. This, by the way, was his frame of mind when he introduced the rabbits onto the smaller island. Rousseau's despair may simply be a sign of weakness of character, but it serves to emphasize that freedom of the will, and the necessity of always choosing that comes with it, places humanity in a difficult predicament.

In grappling with this same predicament, environmental philosophers have alighted on two very different options for bringing freedom into line with nature. The first of these is the speculative path of attributing freedom to nature, seeking to harmonize human activities with a broader cosmic spontaneity. The second is to naturalize freedom, viewing it through the lens of the natural sciences.

NATURE AS FREEDOM

I have already sketched the outlines of a speculative argument for the unity of freedom and nature. In his *Critique of Judgment,* Kant harmonized the two by establishing the possibility of judging nature to be a system created by God with the purpose of supporting the development of humans as moral beings. In the hands of the Romantic nature philosophers, this developed further into the belief that spirit and nature relate to each other not merely as living minds and dead matter but as two different expressions of spirit.

Along these same lines, some environmentalists simply anthropomorphize nature. There are those, for example, who are delighted with the Gaia hypothesis, which seems to hold out the possibility that the Earth is a goddess who can be the object of reverence. James Lovelock proposed

Gaia as testable scientific hypothesis about the interaction of geological, atmospheric, and biological systems on the surface of this planet. According to one commentator, however, almost faster than the hypothesis was proposed, let alone tested, "Gaia had become a unified symbol for pantheism, an acceptably anthropomorphized object for the worship of all that lives."[6] Some elements of this carry over into the "universe story," which draws from both the Gaia hypothesis and modern cosmology and which attributes the evolution of the universe to the unfolding of a single impulse of creative energy.

Academic environmental philosophers who take up the speculative option are generally more circumspect in attributing the characteristics of freedom to the natural world, but the pattern is much the same. The personal, perceptual method of identification used by deep ecologists, for example, depends on the assumption that the self has something in common with the other with which it is to identify. The ethical upshot of this is not a set of moral principles per se but a kind of instinct or impetus to act in defense of organic wholeness and the living things that participate in it, a kind of expansive Self-interest which compels action on behalf of nature: to cut down a tree would be like cutting off a part of oneself. This, rather than a more formal sort of moral obligation, is the basis of what Naess called "biospheric egalitarianism in principle."

Neil Evernden follows suit in his effort to establish a "biology of subjects" in opposition to the Cartesian "biology of objects." He proceeds through a merger of the phenomenological with the biological; as he puts it, "the term 'life-world' may suggest a biological context." Because consciousness is intentional and actively constitutes the lifeworld, Evernden argues, it is impossible to say where the self ends and the world begins. From a biological point of view, by analogy, because humans beings are made up of cells and also because we are parts of larger systems, it is difficult to say where the human body ends and the environment begins. Evernden seems to be claiming that these two sorts of "boundarilessness" may be merged into a single notion of "interrelatedness." Appealing to Maurice Merleau-Ponty's idea of "an organic thought through which the relation of the 'psychic' to the 'physiological' becomes conceivable," Evernden maintains that the self, in a merged psycho-physiological sense, overflows into the surrounding landscape.[7]

The first problem for those who would attribute freedom to nature is that traditional efforts to do so are not only mildly anthropomorphic but anthropocentric as well. I have already mentioned this aspect of speculative nature philosophy: even if Hegel and his compatriots had succeeded, their success would have had consequences that would be distasteful to many environmental philosophers. Kant's critical philosophy and the nature philosophy of the Romantic movement that followed both assumed that the emergence of spirit (i.e., human culture) was the sole purpose of nature, that the order of nature ultimately bowed to the imperatives of spirit. This is firmly beyond the pale of acceptability in environmental philosophy.

In truth, some speculative environmental philosophers find it easy enough to sidestep this problem. All they need to do is to finesse their understanding of nature so that the ends of human culture are subordinate, at least to some degree, to the ends of nature. When Warwick Fox affirms the unity of humans with the rest of the natural world through cosmological identification, he borrows the image of a tree from evolutionary biology to affirm that everything in the cosmos descends from a common source. Implicit in this image is a critique of the great chain of being, the ancient notion that there is a linear hierarchy of beings in which humans stand at the top of earthly creation. Instead, Fox insists, humans are simply one leaf at the end of one twig on the cosmic tree, coequal with birds and bacteria. So while his conception of nature remains to some degree anthropomorphic, humans do not stand at the center of things because the diversity of the cosmos results from the working of a free creative impulse. If Fox could show that human interests are subordinate to the fundamental creativity of the cosmos, he could then begin to construct his case that humans have responsibilities to something larger than themselves.

The second, more serious, problem is not so easily sidestepped. The concept of nature is extremely malleable, and speculative nature philosophy works by selectively reconstructing the concept of nature to serve the objectives of a movement of opposition. This seems to be what is going on with attributions of freedom to nature. Further examination of the work of Rousseau reveals the instability of the Romantic view of nature.

Whether he opted for civil society or for the state of nature, Rousseau's hopes for himself and for humanity depended on the goodness and beneficence of nature and on the guidance humans receive from the "natural" passions of self-preservation and pity. Most often he wrote of nature as divine providence which establishes not only what is but what ought to be; as such, natural law imposes necessary demands and restrictions on human action. Rousseau did not hold back from anthropomorphization. Nature, he insisted, "treats all the animals abandoned to its care with a partiality that seems to show how jealous it is of this right."[8] So long as humans remain in the state of nature, or so long as they approximate it within civil society, nature would take care of them and they would be happy. If this were the whole story, a reconciliation of freedom and nature might well be within reach. It is revealing, however, that Rousseau himself used the term "nature" in at least three other senses that conflict with each other.

First, he often appealed to a materialistic interpretation of nature, according to which the order of things was established not by the grace of God but by the blind working of deterministic laws. There is an element of fatalism in this interpretation, because the workings of natural mechanisms are profoundly indifferent to human efforts to become virtuous. Rousseau was convinced, for example, that regional variations in climate and soil fertility had a decisive influence on human character. So when he considered how a tutor should choose a student, he asserted that the country in which a child lived was not a matter of indifference, for it was only in temperate climes that humans could come to their full growth. Moral freedom and civil liberty could themselves be overridden by natural circumstances: "Since freedom is not the fruit of every climate, it is not within the reach of all peoples." The fertility of the soil would, in large measure, determine what sort of government is appropriate to a given people: the least fertile countries could only be inhabited by savages while the most fertile were suited to monarchies, so that the government could absorb the surplus production and, in effect, keep the people out of trouble. Democracy would only work in small temperate countries in which the soil yielded no more than a modest surplus.[9]

Second, while Rousseau was convinced that nature was good and beneficent only so long as it was untouched by humans, he found support for this conviction in his personal enjoyment of life in the country—amid landscapes which had been cultivated for centuries. When he described landscapes that appealed to him, especially those that were the backdrops for the happiest moments of his life, they were characterized not by primeval forests but by vineyards, orchards, and gardens. The contrast here was not between the human realm and pristine nature but between the country and the city. It did not matter if the landscapes he loved were close to cities so long as they *felt* remote and so long as they offered opportunities for idleness and solitary reverie free from the demands and intrusions of civilized life. Considered thus as country-side, nature seems little more than that which affords a pleasant backdrop. I do not need to dwell on the paradox involved in taking this genteel life as proof of divine provi-dence. One question, however, seems particularly press-ing: What is life in the state of nature supposed to be like? I sometimes picture the "noble savage" ranging across the-French countryside, foraging from vineyards and orchards.

Third, and finally, Rousseau made occasional refer-ences to wildness, a bleak and horrifying nature that he generally chose to ignore. Although his beloved land-scapes displayed the order, harmony, and beauty of na-ture, Rousseau seemed to acknowledge (if only tacitly) that there was another, deeper kind of nature which lay hidden behind the landscape. When he extolled the vir-tues of botany, he described plant life as the "adornment" of the earth, an adornment that was apparently spread over some sort of substrate. He clarified: "Nothing is as sad as the sight of a bare and barren countryside which pre-sents to the eyes nothing but stones, mud and sand." The landscape became appealing to Rousseau's eye only when it had been "brought to life by nature and dressed in her wedding dress in the midst of running waters and the songs of birds."[10] Surely stones and mud are natural, if only in a materialistic sense; a desert is as much a natural landscape as a meadow, and perhaps more so.

What he seems to be up against here is a nature that he could neither understand nor control; one with which he felt no affinity. In a different context, describing his own worries about delays in the publication of *Émile*, he declared that while no misfortune ever really troubled him as long as he knew exactly what it was, he feared darkness and mystery because they were too much the opposite of his own natural openness: "It seems to me that the sight of the most hideous monster would frighten me little, but if I were to glimpse a figure in the night, draped in white, I would be afraid." Even in his own character, supposedly so open and obvious, he encountered that which could not be understood or controlled. He was powerless, for example, to control his anger, his "first involuntary reactions," so he had stopped trying, allowing his blood to boil as it would: "I cede to nature this first explosion, which all my efforts could not prevent or delay."[11] Rousseau had good theoretical reasons to fear wildness. The reconciliation of freedom and nature rested on his assurance that nature was kindly, that he could know that it was looking out for the best interests of humanity. To give explicit recognition to the darker side of nature, one that is both unintelligible and indifferent to humans, might have subverted his entire project.

This is much the same problem that haunted Hegel in his nature philosophy. He was unable to give a satisfactory resolution to the riddle of nature because however much he hoped that spirit might recognize itself in nature, something alien and indifferent remained. The only way Hegel could give the appearance of success was actively to ignore that which is incomprehensible and to insist that reason should have faith in itself. From a skeptical point of view, this is a bad bargain: it seems that humans can ignore that which is incomprehensible only at their peril. In any case, whatever it means for humans to be part of nature, the speculative hope for something like a personal relationship with nature seems ill-founded.

FREEDOM AS NATURE

As I have mentioned, the apparent failure of the speculative path has led some environmental thinkers to draw inspiration and authority from the sciences. The key to this line of argument is the reduction of human freedom to the terms of nature so that ethics is construed as an extension of ecology and evolution. Since the sciences provide the most powerful current accounts of how the natural world works, it might be taken to follow that the scientific study of humanity could yield some insight into human nature. Human ecology in particular might provide a foothold for environmentalists.

Even the earliest practitioners of ecology believed that their new science could be extended to include humans. Stephen A. Forbes, for example, foresaw the possibility of what would come to be called human ecology in his recognition that "man is a part, and in a multitude of cases an all-important part, of the environment of other forms of life," while at the same time "he also is an organism and other organisms are a part of *his* environment."[12]

While it was fairly easy to determine the scope and limits of general ecology once I identified the basic problems and concepts specific to that discipline, human ecology offers no such advantages. Various attempts to develop an ecological study of humans proliferated during the 1920s and 1930s and again in the 1950s and 1960s. They took their shape from a fusion of ecological principles with those of economics, sociology, demography, geography, psychology, ethics, and politics; they met with varying degrees of success. Even so, these attempts have never quite come together as a unified discipline, either in ecology or in the social sciences.[13] Currently, the term "human ecology" does not designate a *scientific* discipline so much as it covers the whole range of environmental concerns that expresses itself as the search for a new "unified vision" of human life on Earth, a search that draws from a wide range of different traditions both within and beyond the sciences.[14]

The scientific study of human life begins, as Forbes suggested, with the deceptively simple observation that humans are organisms. This has two consequences: on the one hand, humans are subject to the influence of various environmental forces; on the other hand, human actions have an impact, often a profound impact, on the ecological systems in which they live. While these two themes of environmental influence and environmental impact have not come together into a single discipline, they have been developed separately by ecologists and others. For the present, I am most concerned with the scope and limits of what can be known about the influence of nature on human nature in order to determine whether human ecology can bypass the problem of freedom.

The belief that human life, character, and culture are shaped by environmental forces is ancient. The medical theories attributed to Hippocrates held that the dominance of one humor over another in the human body could change through the influence of "airs, waters and places," which is to say that climate and geography effect the basic balance of the human body. Because the theory of humors presupposed a close connection between mind and body, humors were believed to determine not only physical health but also character, intelligence, and mental health. By extension, the connection between humors and environmental conditions helped to explain differences in the physical and cultural characteristics of human populations in terms of differences in the climate in which they lived. There was a widespread belief in ancient Greece that peoples that lived in humid climates were less intellectually advanced, for example, because pure, dry air was seen to be the cause of rational thought.[15] This is clearly consistent with Rousseau's belief, many centuries later, that factors such as climate and soil fertility determine whether a given people is suited to democratic government.

It was only in the eighteenth century, in the work of Buffon, that the natural sciences turned to the systematic study of humans taken collectively as a species of animal rather than simply as individuals or populations. Based on then-new observations about the variety to be found among humans around the world, Buffon developed a "grand view" of the unity and diversity of the human species and began to offer natural explanations for some of its features. The unity of the human species was particularly important to Buffon at a time when there was a strong European tendency to divide the species up into "higher" and "lower" categories. He examined those features which distinguish all humans from other species (especially intelligence) as well as the "stages of life," or process of development, through which all humans pass. This did not keep him from a certain European chauvinism, but unlike some of his contemporaries he allowed that even the "savage" races were fully human. In any case, Buffon was the first to offer a natural explanation for the distribution and variety of human races. It is significant that he attributed the variations of human species around the globe primarily to climate.

If Buffon opened the door to a broadly biological treatment of humans, then Charles Darwin walked through it. With the publication of *The Origin of Species* in 1859, Darwin put natural history on a new footing by expanding the scope and power of materialist explanations of natural phenomena. It was not long before human characteristics were examined in this new and harsh light. Just four years after *Origin,* Darwin's friend and chief advocate, T. H. Huxley, published *Evidence as to Man's Place in Nature*. In this his most famous work, Huxley argued on the basis of comparative anatomy that humans are close kin of the "man-like apes," especially the gorilla and the chimpanzee.[16] Darwin expanded the anatomical evidence and added to it behavioral and psychological evidence in *The Descent of Man* and *Expression of the Emotions in Man and Animals,* published in 1871 and 1872 respectively.[17]

It is particularly significant that for Darwin, in contrast to Buffon and his contemporaries, human kinship with other animals includes social instincts, moral sentiments, and intellectual abilities. Both Huxley and Darwin argued that although it seemed as though the differences between humans and other animals were "immense," they were in fact merely differences of degree. In *The Descent of Man,* Darwin devoted considerable attention to the various mental faculties, each of which he traced to an analogue among the "lower" animals. Among these he included language, self-consciousness, individuality, abstraction, general ideas, sense of beauty, belief in God (or in "unseen spiritual agencies"), and the "moral sense." Of this last faculty he wrote: "The following propositions seems to me in a high degree probable—namely, that any animal whatever, endowed with well-marked social instincts, would inevitably acquire a moral sense or conscience, as soon as its intellectual powers had become as well developed, or nearly as well developed, as in man."[18] His point was simply that even the intellect, for so long the hallmark of the separation of humanity from nature, may be accounted for in terms of evolution by natural selection: it is a set of abilities that have conferred a relative reproductive advantage to primates who have been struggling for survival in an unforgiving environment.

As part of this broader movement toward a scientific study of human beings, the first systematic efforts to apply ecological principles to human life developed during the 1920s and 1930s. At the time, many took "human ecology" to be synonymous with the social sciences, due largely to the work of Robert Ezra Park and others of the Chicago school of urban sociology.[19] Park and his colleagues used ecological theory as a model for urban life, on the principle that the city was "a product of nature, and particularly of human nature"; it was a product that was rooted "in the soil."[20] They defined human ecology itself as "a study of the spatial and temporal relations of human beings as affected by the selective, distributive, and accommodative forces of the environment," with particular interest in "the effect of *position,* in both time and space, upon human institutions and human behavior." At root, "the human community has its inception in the traits of human nature and the needs of human beings." For the Chicago school, the city was to be studied with detachment, as a complete system, in hopes of discovering regular and predictable patterns; these patterns were to be regarded as analogous to the patterns of vegetative growth in a prairie community. Borrowing directly from Clements, some proponents of the Chicago school even supposed that human communities undergo a process of succession with predictable periodic "climax" conditions.[21]

The term "human ecology" came back into favor during the 1950s and 1960s, due largely to the emergence of the environmental movement; once again theorists emphasized the application of ecological models to various aspects of human life. In the preface to their definitive collection, *The Subversive Science,* Paul Shepard and Daniel McKinley wrote that "a truly human ecology must be consistent with the broad trans-organic scope of ecology, not merely in an analogical way, but as a real extension." They construed the field broadly enough to encompass an array of very different projects, from studies of human population dynamics using models from animal ecology to cybernetic models of human society.[22]

Given this broad mandate, the success or failure of human ecology depended entirely on the question of how the various sciences relate to one another and how they relate to other domains of human thought. Two disciplines might draw inspiration from the same source (e.g., the organic principle) for very different purposes, and the result is likely to be two very different scientific theories. Because each of the various scientific disciplines is focused on its own unique set of problems, it will develop a set of tools which are customized for its own use, which may not be suitable for work on a different set of problems. To borrow a metaphor from evolutionary biology, a kind of adaptive radiation is at work here: scientific models diverge just as two populations of a single species will diverge when placed in different environmental conditions.[23] If one discipline "borrows" a theory or concept from another, the process of refinement must begin again, and that theory or concept must be changed to fit the new context.

In both of these periods, human ecologists introduced ecological models into the social sciences, but they did not take account of the problem of adaptive radiation. They adapted models of biotic communities and ecosystems from their contexts in ecology, where they used them to answer questions about the organic and physical relationships among living beings. Human ecologists attempted to use these models to answer questions about cultural relationships, such as how knowledge flows through human communities. This is the reason human ecology did not crystallize as a scientific discipline: ecological models can be applied to cultural relationships only metaphorically, in a manner that has neither the same meaning nor the same degree of intellectual authority as ecological models that are applied to biological relationships.[24] The transplantation of a model from one discipline to another can succeed only to the extent that the model is changed to answer the questions particular to that discipline. The resulting model can be judged only on the basis of its usefulness in its new context. It turns out, for example, that while the work of the Chicago school did not result in a unified theory of human nature, it did serve to inspire new research projects in urban geography that had a profound influence on the development of that aspect of sociology.

This same problem arose when Aldo Leopold set out his land ethic. In effect, he attempted to derive ethical consequences from a kind of human ecology: it is a matter of scientific fact that humans are members of biotic communities. Leopold equivocated, however. For his land ethic to work, Leopold had to reconcile two very different meanings of the term 'community': the biotic community is a model for describing physical relationships among living organisms, while the human community is a set of

interpersonal relationships organized in terms of normative prescriptions. Since he did not address this problem
directly, the land ethic rests on the conflation of metaphors taken from various domains in various stages of
refinement. Leopold took up a model which had been
refined by decades of ecological research and debate and
simply merged it with a much broader and much older
sense of community as if they meant the same thing. He
did this, it seems, on the basis of nothing more than a superficial similarity.

Leopold's effort to get around this problem is worth
some attention: he defended the connection between humans and nature by redefining ethics. In ecological terms,
he maintained, an ethic is "a limitation on freedom of action in the struggle for existence," while a philosophical
ethic is "a differentiation of social from anti-social conduct." He assumed that the biotic community and the human community were each a form of symbiosis and collapsed these two definitions of ethics into one: every ethic
is a kind of "community instinct" which helps to assure
the preservation of the community and of the individual.
He thus reduced ethics, the hallmark of freedom, to the
terms of a deterministic instinct. He wrote of philosophical ethics as undergoing a natural process of evolution
through which ethical consideration was extended to a
wider and wider circle of beings.[25]

In defense of Leopold, Callicott develops this idea of a
community instinct by way of Hume and Darwin. Hume
understood ethics in terms of moral sentiments—"fellow
feeling, sympathy, benevolence, affection, generosity"—
which are fixed characteristics of human nature; they are,
in other words, matters of psychological fact. Darwin provided the plausible explanation that these sentiments came
to be fixed by natural selection because membership in a

coherent and stable community had a survival advantage for the ancestors of Homo sapiens. Like Leopold, Callicott attempts to gain a foothold for environmental ethics by construing "coherent and stable community" broadly: moral sentiments need not be limited to one's tribe, or even to one's species. For the other species, it is only a matter of what the individual regards as the community which is to be the object of affection. By making it possible to perceive the natural world as a living whole, Callicott argues, "ecology and the environmental sciences . . . inform us of the existence of something which is a proper object of one of our most fundamental moral passions." In other words, the moral sentiment of which society as a whole is the proper object can be extended to embrace ecological systems as well.[26]

Insofar as he reduces ethics to a matter of social instinct, Callicott faces a serious paradox. Recall that the goal of this kind of environmental philosophy is to provide a moral context for human life, one which places a set of obligations or restrictions on human activity. If freedom confers moral significance on human actions, and if freedom is a prerequisite for meaningful moral debate, then an environmental ethic that undermines freedom effectively undermines itself. More specifically, if Callicott's argument works, then there is no point in his offering any kind of argument at all: people either have the appropriate instincts or they do not. For that matter, there is no standard according to which instincts can been deemed appropriate or inappropriate other than the fact that people actually have them. Within the framework of a theory of moral sentiments, after all, self-preservation and sexual appetite are on exactly the same moral plane as the social sentiments of which Leopold and Callicott approve: they are all perfectly natural, and they all contribute something to the survival of the species.

The broader point here concerns the limits of the sciences more generally. Since Darwin's time, the natural sciences have been delving ever more deeply into human nature by way of physiology, neurology, psychopharmacology, genetics, and so on. It is seldom acknowledged, however, that models of various aspects of human nature are no more adequate than models of other natural phenomena, and there is no reason to believe that any model or combination of models will tell the whole story of human life. It is reasonable to conclude that human life can be considered from any number of points of view, none of which is exclusive of the others.

In the end, neither the speculative nor the scientific strategy seem to provide a clear and unambiguous vision of the place of humans in the natural order as a whole. The problem of freedom raises serious doubts about the prospect for an ethically motivated integration of humans into nature, even when one sets aside all of the problems involved in determining what nature is. Indeed, I have come to doubt that there can be a tidy resolution to the problem of freedom and nature that will define, once and for all, our proper relationship with our environment. It may well be the case that human life is beset by a fundamental ambiguity. We must choose from among a wide range of options without much hope for sure and steady guidance about which options are best. Yet our lives and activities are shaped by forces beyond our control, including some that have been set in motion by our own choices and actions. As we pursue our various projects, we change the world around us, but we are always vulnerable to the material and cultural consequences of those changes. In practical terms, this suggests that there is good reason to be extremely cautious when deciding how to live and what to do, since there is no guarantee that everything will work out for the best.

PLACE

There are those who would argue that everything I have considered up to this point has been far too abstract—too caught up in the global relationship of freedom to nature—to do much good anyway. If we want to find our proper relationship with our environment, they argue, we must look closer to home. For instance, bioregionalists maintain that people should think of themselves not only as members of biotic communities in general but also as members of *particular* biotic communities: to them, the details of local life are of irreducible importance in deciding how to live our lives. Kirkpatrick Sale describes this as a matter of scale. People cannot connect with and are not moved by abstract, theoretical arguments about global-scale environmental problems, he argues. "The only way people will apply 'right behavior' and behave in a responsible way is if they have been persuaded to see the problem concretely and to understand their own connections to it directly—and this can be done only at a limited scale." The appropriate scale, he maintains, is the bioregion and, within the bioregion, the community. Bioregions are the "natural" regions of the earth, defined by the contours of the landscape, which are "not hard to find by using a little ecological knowledge."[27]

Ecological details aside, some have argued on behalf of place as such as a crucial element in any understanding of human life in the world. To understand this argument, I must draw out one more corollary of Cartesian dualism.

For Descartes, material substance is characterized by extension: a material object has measurable dimensions of height, width, and depth. The proximity of any two material objects can be expressed in terms of such measurable dimensions (e.g., the cue ball is two feet away from the eight ball). Space is the abstract, uniform expanse in which material objects have their existence and their relations to one another; it is the field of pure dimension.

Place, on the other hand, is anything but an abstract uniform expanse. Rather, it is the surrounding world with which a person is most directly involved and which is invested with meaning. In *Getting Back into Place,* a work drawn from both the phenomenological tradition and his own experience, Edward Casey seeks to establish a renewed respect for place in this sense "by specifying its power to direct and stabilize us, to memorialize and identify us, to tell us who and what we are in terms of *where we are* (as well as where we are *not*)." This defining power of place is important: "Where you are right now is not a matter of indifference but affects the kind of person you are, what you have been doing in the past, even what you will be doing in the future. Your locus deeply influences what you perceive and what you expect to be the case."[28]

There is something truly appealing about the small-scale approach. At the practical level, at least, it is clearly important to pay attention to immediate experience and to local detail. Sale and others may well be correct that it is only at the local level that people engage with what is generally called "the environment"—in the primary sense, the surrounding world with which one is concerned on a daily basis. This might account for the difficulty of mobilizing people to do something about global warming: the problem is simply too abstract and too remote to present itself as a matter of immediate concern. It is only when some local environmental change can be linked to the larger problem that people begin to take notice of it.

As a theoretical move within environmental philosophy, the turn toward place also makes good sense. It seems, at least in principle, to leave open the possibility of a more nuanced relationship between freedom and nature because it can avoid both the oceanic holism of deep ecology and the dilemma of moral naturalism. Also, more practically, it might open the door for philosophers to involve themselves more directly in the environmental problems of their own communities.

The key question for my purposes is whether this attention to place *as such* or to the ecological details of any given place can "direct and stabilize" human activities in a way that might satisfy a speculative environmentalist. Will it, in other words, somehow make clear to us what are the "natural" limits on human life? This is what Sale has in mind: what we discover in the bioregion is nothing other than the local manifestation of "Gaea's laws."[29] Some phenomenologies of place also point toward a better, more "authentic" mode of dwelling in place; at the very least, they begin with a sharp-edged critique of the displacement brought about by the advance of science and technology.

I can point out two limitations to an environmental ethic of place. First, the problem of freedom and nature remains, and it does not seem as though shifting our perspective from the global to the local can change that. The problem takes on a more specific form, however, when one considers the degree to which places are constituted (that is, given meaning) by humans. Casey has noted that "implacement is an ongoing cultural process with an experimental edge. It acculturates whatever ingredients it borrows from the natural world, whether these ingredients are bodies or landscapes or ordinary 'things.'" As a consequence, places cannot be seen as "natural" in any straightforward sense, and some account must be given of the intertwining of the natural and the cultural in the experience of place.[30] If Casey is correct, then even a bioregion is not really defined by nature: some humans have picked out this or that aspect of the landscape—watersheds, for instance—and given it significance for what may be a wide array of reasons.

Second, an ethic that is bound to a particular place, and to the culture of that place, may not be applicable to other places and cultures. In other words, the turn toward place can be seen to entail a radical kind of pluralism in which the practices and principles of one culture are incommensurable with those of another and in which communication between the two cultures is difficult at best. Deane Curtin has examined cross-cultural disputes of this type. He concludes that while the moral world of a culture might be transformed over time, it cannot be reduced to that of another or made subject to some abstract set of rules. Instead, a moral world has "objective positional integrity." This positional objectivity is not at all to be equated with truth; it is simply "the best account of internal goods from within a practice."[31] So even if an ethical tradition is rich and well-established in its place, it is not necessarily translatable to other contexts, and it does not necessarily guarantee the long-term well-being of either the place or its inhabitants.

[4]

The Moral Compass

However difficult it might be to describe—or prescribe—
the place of humans in nature, environmental philosophers continue their search for a moral compass to guide
humanity out of the environmental crisis. Aldo Leopold's
land ethic serves as the prototype. He envisioned a historical and evolutionary process by which human ethics
become more and more inclusive over time. He predicted
that this process would culminate in an extension of ethics to encompass the non-human world. Any system of
ethics, he maintained, is based on the principle "that the
individual is a member of a community of interdependent
parts." Leopold redefined ethics so that it can apply to the
biotic community as well as the human community. The
upshot of this is that when ethical consideration passes beyond the boundaries of the human community, decisions
about environmental change can no longer be made solely

on the basis of expediency. It becomes necessary instead to "examine each question in terms of what is ethically and esthetically right, as well as what is economically expedient. A thing is right when it tends to preserve the integrity, stability and beauty of the biotic community. It is wrong when it tends otherwise."[1]

With this dictum, Leopold set the tone for traditional environmental ethics, which was first established as an academic discipline in the early 1970s. Environmental ethicists took up the task of formulating, refining, and defending moral rules that place demands or restrictions on any human activities that alter the natural environment. Although they adopt a wide range of approaches, they generally agree that there is a need for a specifically *environmental* ethics, a set of normative claims that can govern human interactions with the surrounding world. This, they believe, sets their work apart from most of the ethical tradition: principles formulated to govern relationships among humans are not adequate to the task of generating the kind of demands and prohibitions on human activity that are compatible with environmentalist goals.

Some go farther than this, claiming that the anthropocentric bias in traditional ethical theories makes them complicit in destructive practices. If it is the case that humans alone have intrinsic value, then non-human entities and systems can have value only insofar as they serve human interests; any value they have is merely instrumental, not intrinsic. As a consequence, human treatment of non-human entities is usually a matter of expediency. Suppose I destroy a mechanical clock. In doing so, according to this account, I have not harmed the clock in any morally significant sense; it is not the sort of thing that can have an interest in its own welfare. However, I may have harmed the owner of the clock, at least indirectly.

So it may be that I had an indirect duty not to destroy the clock as an extension of my direct duty not to harm the interests of another human being. Of course, if I happen to be the owner of the clock, or if the owner explicitly asked me to destroy the clock for some reason, then there is no ethical problem at all. In the Cartesian universe, the same reasoning would apply to living organisms and ecological systems as well. With this reasoning, critics charge, traditional ethics has condoned—tacitly or explicitly—all manner of environmental degradation.

To counter this bias, environmental ethicists propose a variety of non-anthropocentric principles. Their proposals entail a shift in emphasis at both levels of ethical theory: meta-ethics and normative ethics. At the meta-ethical level, ethicists in general are concerned with the framework within which ethical debate takes place, including the nature of the evidence that can support normative claims. Environmental ethicists, for their part, hope to establish the possibility of attributing intrinsic value to non-human entities and systems. If an entity has intrinsic value, moral respect or concern for that entity must be taken into account, and it cannot be reduced to mere instrumental or economic value to be traded off in favor of some other instrumental or economic value.

Among environmental ethicists, however, there are several points of disagreement regarding the nature of intrinsic value. Is intrinsic value an objective property of natural objects that can exist independently of the existence of valuers, or is it a subjective or relational property that arises only when humans or other valuers acknowledge it? Consider one example. In his attack on the practice of ecological restoration, Robert Elliot develops an indexical theory of intrinsic value, according to which a thing has intrinsic value if it stands in a relation of approval to a given attitudinal framework. In other words, intrinsic value is always indexed to the attitudes of beings that are capable of approval or disapproval. So if I approve of the naturalness that is exhibited by a particular landscape in a manner that is consistent with my broader set of values, then the landscape has intrinsic value for me.

Elliot offers this theory in direct opposition to the non-naturalist objectivism of G. E. Moore. While Moore claims that intrinsic value is an unanalyzable non-natural property of a thing, a property to be known only through a kind of moral intuition, Elliot maintains that the value-adding properties of things, those that can meet with approval, are natural properties that can be known by more ordinary means. The advantage of this approach, Elliot believes, is that it makes possible a much more interesting kind of normative debate: my moral judgment can be corrected if, for example, I turn out to be wrong in attributing naturalness to the landscape in question.[2]

Environmental ethicists also disagree about the relationship of meta-ethical theory to matters of environmental policy. Monists insist that environmental ethics must speak with a single voice and that all of the values it fosters must be grounded in and informed by a single coherent worldview. While J. Baird Callicott allows for a "multiplicity of hierarchically ordered and variously 'textured' moral relationships," for example, he insists that to reject a single metaphysic of morals is to flirt with nihilism.[3]

Pluralists, on the other hand, argue that environmental ethics encompasses not only a multiplicity of values but also a multiplicity of worldviews and meta-ethical theories. Bryan Norton's defense of pluralism is essentially strategic: if the task of environmental philosophy is to articulate a range of values, pluralism is simply the best way to encompass them all and the most likely path to good environmental decision-making. His argument is based largely on an understanding of the relation of philosophy to practice that is quite different from that of most environmental ethicists, including Callicott.[4]

Normative ethics, the second level of ethical theory, is concerned with the formulation and defense of specific normative claims that can serve as guides to conduct. A number of environmental ethicists have adopted a biocentric position at this level; they insist that every living being deserves moral consideration. One line of argument follows from Kant's contention that a being with free will is an end in itself and so has intrinsic value. Non-human animals—as well as plants, fungi, bacteria, and so on—may not possess free will as such, but they do engage in goal-directed activity. Organisms are, in the words of Paul Taylor, "teleological centers of life,"[5] and as such can all be construed as ends in themselves in the sense that they have goals and interests of their own. In an important synthesis of this branch of environmental ethics, Holmes Rolston maintains that organisms are normative systems, which is to say that organisms have a "nature" that determines not only what they are but also what they ought to be and that this normativity has a claim on human respect.[6]

Another option for normative environmental ethics is to follow Leopold in shifting the focus of moral concern from individual organisms to larger ecological systems. Usually known as ecocentrism, this option is closest to what I have constructed as the central case of environmental philosophy. As a consequence, the outlines of an argument for an ecocentric ethic should be familiar by now. In brief, ecocentrists generally hold that a holistic interpretation of ecosystems yields two distinct but interconnected consequences for ethics.

The first consequence is that ecological systems can be seen as analogous to organisms: they function by a complex network of internal relations that tends to maintain the system in a state of balance. As with the health of organisms, the equilibrium proper to a given system has value and ought to be respected for its own sake. Holmes Rolston draws explicitly on the organism-ecosystem analogy when he extends his biocentric theory to the larger systems of nature. The second ethical consequence is that once people come to recognize that they are part of a larger system, they can conceive of or experience both their participation in that system and their affinity with the other living organisms that participate along with them. On the basis of this experience, they will be motivated to protect and preserve the system for its own sake.

Before I move on, I should note two other wellsprings of diversity within normative environmental ethics. The first concerns the relationship between values and obligations. Once I have acknowledged the intrinsic value of a natural entity, what are my duties toward it? What ought—or ought not—I to do with respect to that entity? Environmental ethicists have drawn from the full range of traditional answers to this question—especially consequentialism, deontology, and virtue ethics—and have added variations of their own. Consider the case of animal rights. Classical utilitarianism holds that pleasure has intrinsic value, from which is supposed to follow the obligation to maximize pleasure and minimize pain in considering the consequences of action. Peter Singer has argued that since animals are capable of suffering and because people have an obligation to minimize suffering, then people generally ought not to harm animals. The emphasis on the intrinsic value of animals themselves lends itself more to the deontological argument that whatever their consequences, actions that affect animals ought to be motivated by respect for their status as teleological centers of life. Finally, virtue ethics concerns not so much the consequences or motives of action as the character of those that act: Do I want to be the kind of person that abuses animals?[7]

Environmental ethicists have also found a variety of ways to deal with the difficult problem of balancing human interests with those of non-human entities and systems. Few would argue that either biocentrism or ecocentrism implies that human interests are to be ignored altogether. The point, usually, is simply that there are a number of values in contention and that the intrinsic value of non-humans should not be trampled on the way to fulfilling the economic self-interest of human beings. In this spirit, Leopold exhorted his readership to "examine each question in terms of what is ethically and esthetically right, *as well as* what is economically expedient."[8] He did not dismiss economic interests; he simply insisted that they should not be the sole basis on which decisions are made. Even with this proviso, two questions remain: How should the balance of interests be struck? and Under what circumstances may the interests of human beings be said to outweigh the interests of non-human beings and systems? These questions have led many environmental thinkers to formulate priority principles which make the balance of interests explicit.

Beyond environmental ethics, in the broader domain of environmental philosophy there is an even greater variety of ethical possibilities. A number of environmental thinkers have expressed their dissatisfaction with the tools and practices of traditional philosophical ethics and so have sought out different ways of conceiving or perceiving the ethical dimension of the human-nature relationship. Some, including deep ecologists, even see danger in the very act of formulating moral principles. According to Warwick Fox, such principles "are directed to and thereby reinforce the primary reality of the narrow, atomistic or particle-like volitional self." In other words, ethics seeks to regulate the relationship between the self and others, but because the self-other distinction has been dissolved by deep ecology, "ethics (conceived as being concerned with moral 'oughts') is rendered superfluous!" Instead, deep ecologists propose a much more direct intuition of concern and responsibility. Arne Naess claims that "the ecological field-worker acquires a deep-seated respect, or even veneration, for ways and forms of life"; this is a matter of intuition, a direct grasp of responsibility. Fox places this intuition in the context of identification: when one has realized the wider self, "one will naturally (i.e., spontaneously) protect the natural (spontaneous) unfolding of this expansive self (the ecosphere, the cosmos) in all its aspects."[9]

Opposition to traditional ethics is a rare point of agreement between deep ecologists and ecofeminists. Karen Warren proposes a "reconception of what it means to be human, and in what human ethical behavior consists." Eliminating the logic of domination involves

> a shift *from* a conception of ethics as primarily a matter of rights, rules, or principles predetermined and applied in specific cases to entities viewed as competitors in the contest of moral standing, *to* a conception of ethics as growing out of what Jim Cheney calls 'defining relationships,' i.e., relationships conceived in some sense as defining who one is.

The result is a contextualist and inclusivist ethic that "makes a central place for values of care, love, friendship, trust, and appropriate reciprocity."[10] The establishment of an "ethic of care" is a common theme in ecofeminist literature.

From a skeptical point of view, there are at least two reasons to doubt that environmental philosophers of any stripe can provide a moral compass specifically suited for environmental decision-making. The first concerns the relationship between facts and values. Even if it is possible to have reliable knowledge of nature and of the human relationship with nature, it does not necessarily follow that we can derive moral obligations from that knowledge. The second reason concerns the standards and practices adopted by many environmental philosophers, especially environmental ethicists. The reactive character of environmental speculation carries over into the ethical domain as well; selective philosophy and selective science are extended and reinforced by the appeal to "adequacy" as a kind of political litmus test for ethical theory.

FACTS AND VALUES

The problem of the relationship between facts and values arises because environmental philosophers must necessarily appeal to some sort of description of the natural world to inform their ethical standards, and they often appeal to descriptions provided by the sciences. Even those who do not formulate a complete ecological worldview must at least describe the entities and systems to which they attribute intrinsic value. The problem is that, as Hume pointed out, a moral prescription (an "ought") can never be derived logically from an empirical description (an "is"); facts and values must be established on very different bases. Environmental philosophers have used various tactics for overcoming this restriction so that their ethical injunctions or intuitions can in some sense be grounded in ecology. In effect, the entire development of environmental ethics can be read as a series of ever-more-sophisticated attempts to overcome the is/ought dichotomy, to link facts and values by some means other than logical deduction.

In Chapter 3, I discussed Leopold's redefinition of ethics in ecological terms and Callicott's subsequent appeal to a Darwin-inspired theory of social instincts. Now I can make it clear that both of these tactics are intended to bridge the is/ought gap. If moral sentiments are naturally occurring phenomena, then it makes sense to suppose that they can be shaped in some way by outside forces, including an individual's understanding of how the world works. If, as Callicott argues, humans naturally tend to identify with a larger community, then all that is needed to establish an environmental ethic is an appropriate understanding of what the relevant community is. Ecology leads to a recognition of the biotic communities of which humans are part, from which it follows that humans will tend to identify with biotic communities. So facts shape values. The new understanding of the facts brought about by ecology does not change the moral sentiments themselves, however; it simply stirs them in novel ways.[11]

Another way to bridge the is/ought gap is to claim that facts and values are always already intertwined. Rolston, for example, states boldly that "one's beliefs about nature, which are based upon but exceed biological and ecological science, have everything to do with beliefs about duty. The way the world *is* informs the way it *ought to be*." Description and evaluation of nature are intertwined in what he calls "metaecology," through which "an *ought* is not so much *derived* from an *is* as discovered simultaneously with it." He goes on: "For some, at least, the sharp *is/ought* dichotomy is gone; the values seem to be there as soon as the facts are fully in and both alike seem properties of the system." In an argument that finds echoes both in the universe story and in Fox's transpersonal ecology, Rolston cites a trend in nature: "its projecting of life, stability, integrity, culminating in a sense of beauty when humans enter the scene."[12]

Using a similar tactic, some environmental philosophers have appealed to phenomenology to provide a perceptual bypass of the is/ought dichotomy. Within the more traditional approach to environmental ethics, for example, Don Marietta has used phenomenology to foster a recognition that values are embedded in all human experience. He supposes it to follow that there is no need to derive an is from an ought because the two are always discovered together. Even the sciences, Marietta argues, are shot through with values. As I noted in Chapter 2, he lists words used in ecology that are also used to express environmental values: stability, diversity, unity, balance, integrity, order and health. He regards this not as a coincidence but as the reflection of a fusion of fact and value in human experience, from which both ecology and environmental ethics spring.[13]

Beyond the standards and practices of traditional ethics, more radical efforts to articulate the relationship between fact and value are often based on an appeal to the works of Martin Heidegger, which have long been seen as a rich resource for environmental thought. Bill Devall and George Sessions point out that Heidegger made three important contributions to the literature of the deep ecology movement, the third of which is that he "called us to dwell authentically on this Earth, parallel to [deep ecology's] call to dwell in our bioregion and to dwell with alertness to the natural processes." Toward the end of his life, they maintain, Heidegger "arrived at a biocentric position in which humans would 'let things be.'"[14]

Along these lines, Bruce Foltz has offered a reading of Heidegger that culminates in a radical revision of the meaning and import of environmental ethics. Foltz proceeds by distinguishing two senses of 'environment.' Specifically, he writes, "that which is presupposed and employed by the biological sciences, including ecology, is not, strictly speaking, an environment at all—not a surrounding world, an *Umwelt*"; it is, instead, "the given 'environ' . . . of plants and animals, of those entities that 'have' no world." The 'environ' is thus the material surroundings of an organism and the material relations that organism maintains with other material beings. By contrast, the "true environment" which is to be the basis of his "genuine environmental ethic" is precisely the *Umwelt:* "that which we inhabit most immediately, that which concerns us and matters to us most persistently, and hence that whose meaningfulness intertwines most continually with the course of our lives."[15]

This redefinition of environment gets its ethical force, Foltz believes, from Heidegger's notion of "dwelling poetically" upon "the earth." Foltz unpacks this notion as follows:

> That which genuinely surrounds, a true environment, is disclosed, established, and conserved only by the poetic comportment that concerns itself in an attuned manner with entities— concerns itself that they be conserved within what is essential to them—and that thereby allows them to be what they are, to matter and be significant, to be near.[16]

The key Heideggerian terms to which Foltz appeals bear a striking if superficial resemblance to the jargon of contemporary environmentalism: "the conserving that is fundamental to dwelling involves saving the earth."[17] For those who are looking for guidance and who are already inclined

to support environmentalist goals, "dwelling," "conserving" and "saving the earth" can suggest that particular
ways of living, particular policies, even particular things,
are preferable to others. In other words, it may seem that
to dwell poetically is to live sustainably.

There are two serious problems with all of these various efforts to overcome Hume's restriction, the first of
which is that the facts are not all that clear. I have tried
to show that there is no particular reason to believe that
the human mind has access to the true nature of nature,
whether through speculation, perceptual immersion, or
scientific inquiry. Appeals to ecology to reveal some fundamental integrity and stability are not likely to be satis-
fied. The appeal to "metaecology" is no help, either, as it
is simply a conflation of science with speculation that is
subject to the same kind of doubts.

The more serious problem is that even if the facts are
clear, the values that accompany them remain ambiguous
and subject to dispute. I do not question that facts and
values are difficult—if not impossible—to separate in human experience, and I accept that the relationship between them may be something subtler than the kind of
logical deduction forbidden by Hume. On the other hand,
I would argue that Hume is basically right in his belief that
facts and obligations are at least relatively independent
of one another, so that the validity of a normative claim
must be established, if at all, on grounds that are essentially different than those of the facts with which they are
associated.

Many environmental ethicists formulate their arguments in the hope that the success and authority of the
sciences can somehow rub off on ecologically inspired
values. As an example, consider the use of apparently
value-laden terms in ecology. If scientists have reason to

agree on the use of value-laden terms to describe the facts in their domain, Marietta and others have argued, then there must also be reason to agree on the values with which the terms are laden. I would argue, to the contrary, that even if the sciences could establish a set of facts beyond doubt, and even if the facts were to be described in terms that are reminiscent of certain value judgments, it does not follow that the sciences can lend any of their authority to any particular set of values or obligations. Rolston's appeal to "metaecology" and Naess's appeal to "ecosophy," do not strengthen the fact-value link; they weaken it considerably by rendering it more transparently arbitrary.

To clarify the role of values in the sciences, it will be helpful to begin with a case study. In Chapter 3, I examined one part of the domain that has come to be called human ecology; namely, the study of the influence of the natural environment on human life and character. Another aspect of human ecology examines and evaluates the impact of human activities on nature. These two fields of study—influence and impact—form the two halves of a line of inquiry that is vital if we are to understand and cope with environmental problems. We humans change our environment, but since we are ourselves products of our environment, we are vulnerable to the consequences of those changes. I do not question the urgent necessity of understanding the physical processes with which we are intertwined and upon which we depend; given our vulnerability, willful ignorance is likely to be fatal. My concern is whether the scientific study of the impact of human activities can also support moral judgments about those activities.

According to Donald Worster, natural history and, later, ecology have accommodated two very different normative stances toward human-induced environmental change; he labels these the "arcadian" and the "imperialist" traditions, respectively. As Worster portrays it, the arcadian tradition corresponds roughly to a worldview based on the organic principle whereby humans are part of a larger organic whole that they are obliged to respect. He maintains that the arcadian tradition found its first "ecological" expression in the work of the eighteenth-century naturalist Gilbert White, who explored the rural world he inhabited and so discovered an "arcadian harmony with nature."[18]

The work of George Perkins Marsh in the mid-nineteenth century is typical of this tradition. Influenced by Humboldt and others, Marsh assumed that nature is in a delicate and harmonious balance. He assumed further that humans are essentially a destructive force. He wrote:

> Apart from the hostile influence of man, the organic and the inorganic world are . . . bound together by such mutual relations and adaptations as secure, if not the absolute permanence and equilibrium of both, a long continuance of the established conditions of each at any time and place, or at least, a very slow and gradual succession of changes in those conditions. But man is everywhere a disturbing agent.[19]

Nevertheless, Marsh had some faith in the abilities of humans to do the right thing. Since he attributed the destruction of natural harmony to human ignorance and negligence, for him there was always the possibility of restoration and preservation through careful and attentive management. His book was, at least in part, a moral exhortation to action.

The imperialist tradition, as Worster portrays it, adheres to a crude version of the mechanical principle, according to which humans are the managers of nature with divine authority to manipulate natural systems to serve human interests. This outlook has its origins in the work of Bacon and Descartes. According to Worster, the imperialist tradition was given its first "ecological" expression by Linnaeus. In Linnaeus's work, the doctrines of mechanism and divine providence seem to grant humankind "explicit permission to manage the natural economy for its own profit."[20]

Buffon, whom I have already credited with revolutionizing natural history, might also be identified as a strong spokesman for the imperialist tradition. At the heart of his work on the "natural history of man" lay a managerial ethic which directed Buffon's attention and gave him a basis for interpreting what he discovered. While Buffon acknowledged that there is a certain orderliness to nature and that humans do sometimes harm that order, he believed nevertheless that nature, left to itself, is "brute and hideous and dying." Only human cultivation could revive nature and make it agreeable and suitable for human use. Buffon was not so arrogant as to assume that humans had any definitive or secure dominion in the world: the human right to cultivate nature was not bestowed by divine providence but had to be seized by conquest. So Buffon seemed to cast humans in the role of insecure conquerors of a world that was not entirely hospitable to them.[21]

Worster and many others would cast their lot with the values embodied by the arcadian tradition and would like to have some sort of historical or scientific backing for their preference. But it would seem that the only way to secure this backing would be to practice selective science: assert that Marsh was essentially right and Buffon was essentially wrong and cite the work of this or that latter-day scientist to support the contention. This is at best a dubious enterprise.

Even the arcadian/imperialist split is deceptive. Worster introduces it in the context of a history of "ecological" ideas, but it is clear enough that he is not content to study the development of ecology as a scientific discipline. Rather, he offers a partisan history of the origins of "ecological" thought in a broader sense, encompassing the whole array of ideas, from whatever source, that have come to inform contemporary environmentalism. If in the process he can construe the development of at least some aspects of ecology (the science) as supporting ecology (the movement), then so much the better. When he does examine the work of natural scientists, he seems mostly to be concerned with picking out friends and foes based on the values that seem to be embedded in their work.

Unfortunately for Worster, the natural sciences cannot easily be divided into friends and foes. A more skeptical, less selective reading of the history reveals that neither the arcadian nor the imperialist tradition of moral judgment has won out in the development of ecology as a science. Even if one of them had, a more skeptical view of the scope of scientific knowledge would limit the kind of conclusions that could be drawn from that fact. Suppose that the research agenda initiated by Marsh had come to single-handedly dominate the environmental sciences. At first glance, that might seem to validate Marsh's ethical assessment of human activities. Even if the sciences are necessarily informed by the values of scientists—and there is plenty of evidence to suggest that they are—it does not follow that scientific research can then lend its authority to judgments of value. Let me explain this in more detail.

The sciences are permeated with values of all kinds. At the very least, a number of intellectual or methodological values are in play—and often in dispute—in the work of scientists. There would be no scientific inquiry at all if scientists did not choose, for example, whether a given bit of evidence is important, whether a given theory is preferable to its competitors, or whether a given research project is worth pursuing. At least some of these decisions will be based on a set of values that are instilled by education in the sciences, such as predictive power, empirical content, or elegance. Still more broadly, the work of scientists presupposes, tacitly or explicitly, that scientific knowledge itself is worth pursuing.

Even if scientists all agreed that the ultimate goal of their work was to make possible the prediction and control of natural phenomena, the decision to pursue this kind of power in a particular domain of human experience is itself a question of values. Scientific research programs are often motivated by the need to solve particular practical problems in order to produce or preserve various kinds of social values. So, for example, the early development of the study of population dynamics in ecology was motivated by an interest in developing techniques for the biological control of pests that would serve the ends of agriculture and, ultimately, human well-being. The idea of biological, as opposed to chemical or mechanical, control is to limit the abundance and distribution of the pest species by introducing a specialized predator into the biotic community. The pressing need to be able to predict the consequences of such an introduction led to the development of more sophisticated mathematical tools in population ecology, especially those concerned with predator-prey interactions, which contributed to the general movement toward the use of mathematics in ecology.[22]

Likewise, in the 1950s, the need to assess the effects
of human-generated ionizing radiation on natural systems
provided the impetus for the rapid development of ecosystem ecology. Here political institutions played an important role: a feedback loop developed between ecology and
nuclear energy in the decades following World War II as
the Atomic Energy Commission became an important
source of funding for ecology and some of the problems
associated with nuclear power provided a justification for
ecosystem research.[23] Partly as a result of these practical and economic considerations, radioactive isotopes became an important tool for "pure" ecological research.
The Odum brothers' work on a coral reef on Eniwetok
Atoll, one of the nuclear weapons test sites of the United
States, was the first important example of this research.
By determining where and when radioactive isotopes ap-
peared in various components of the local ecosystem, the
Odum brothers were able to trace in detail the pathways
of the reef's "metabolism."[24]

Because scientific inquiry is a human enterprise, it is
motivated and informed by other values as well. These
might range from broad social goals to the priorities built
into institutions to the priorities built into theoretical
frameworks to the idiosyncratic preferences of each particular scientist. Arrogance, greed, lust for power, simple
curiosity, hope for liberation, charity, and humility might
all be in play, alongside countless other values. In the end,
the sciences are not all that different from speculation.
Scientists, like the natural philosophers before them, develop models of the world that are shot through with their
own personal values. It should be neither surprising nor
offensive that Buffon and Marsh each offer explicit support for a strong ethical stance toward human activity in
the natural world.

Nevertheless, the power of the sciences lies in the procedure whereby idiosyncratic views are opened up to critical scrutiny. This procedure selects and refines those concepts and theories that can contribute to the prediction of natural phenomena. If the value-laden perspective of a particular scientist turns out to be useful for this purpose, then those values may come to be embodied in a set of heuristic principles that inform further research and debate; as such, they tend to lose their status as ethical principles as they gain status as scientific tools. Even so, scientific inquiry is an open-ended process that is moved forward by a return to speculation, which projects human visions and hopes onto the world so that values are continually re-introduced.

If the arcadian and imperialist traditions really did exist, as Worster claims, then they would have been radically transformed as ecology developed as a scientific discipline in the twentieth century, as would the organic and mechanistic metaphors with which they are associated. Ecologists inherited a much more mundane version of the organic principle than the one the arcadians and the Romantics embraced. In its secular form, the organic principle had already been transformed through its interaction with mechanistic accounts in the development of physiology. As a consequence, ecology has been concerned not with the balance of nature in some abstract cosmic sense but with the relative stability of this or that ecological system, studied at this or that spatial or temporal scale.

Ecologists have gradually brought every aspect of the organic principle under scientific scrutiny, including, most recently, the idea of equilibrium itself. Let me draw out the ethical consequences of this idea. Non-equilibrium theories in ecology complicate matters for environmental philosophers and further attenuate any moral values that are supposed to be derived from or discovered alongside ecological research. If it is not possible to establish an objective standard for ecological health (i.e., what an ecosystem ought to be), then it is much more difficult to argue that humans ought to preserve and protect the health of ecosystems.

So values are introduced into scientific research at every turn, but to draw out values from this research would be to argue in a circle. It does happen that scientists will, supposedly on the basis of their research, pass moral judgment on this or that kind of human activity. Even if their research was incontrovertible, it would not follow that the associated ethical judgment is more valid than any other. Even if ecologists were unanimous and unwavering in their support of a particular kind of biocentrism, their status as scientists would contribute nothing to their advocacy of that standard. If I find support in the sciences for the arcadian view, it is only because I, or the scientists to whose work I am drawn, or both, have presupposed the positive value of untrammeled nature and the negative value of human alterations. Marsh was looking for evidence that humans have disturbed nature; it should hardly come as a surprise that he found it. Nor should it be a surprise that advocates of the imperialist view find evidence that nature is incomplete without human intervention. As a historian, Worster is looking for evidence of a Manichean clash between two worldviews; he, too, finds what he is looking for.

In case this seems like nothing more than an attack on the arcadian view, I should note that the skeptical assessment of the problem of facts and values does not leave the imperialist tradition unscathed. The managerial ethic is characterized, above all, by its confidence in human

knowledge and human abilities and its faith that everything will work out for the best. If anything, the development of ecology has served only to undermine this confidence. Ecological systems are more complex—and
adequate models more elusive—than Linnaeus, Forbes, or
Clements could have imagined. Even if new ecological
models can provide enough predictive power to make
large-scale management feasible, the advancement of scientific inquiry into various aspects of human life should
dispel any imperialist illusions that nature exists solely in
order to serve human interests, to assure human well-
being, or even to assure human survival. Even though
humans must, out of necessity, make use of the surrounding world, this is not a matter of divine right: plants and
animals have interests of their own, material resources
are either available or unavailable, and there is no guarantee that even the best-laid plan will succeed. In this regard, Buffon saw matters more clearly than others to
whom Worster attributes the imperialist view, including
Linnaeus.

So much for deriving values directly from the sciences. The other main path toward the fusion of facts and
values, what I have called the perceptual bypass, does not
work either, for similar reasons. In general, the analysis of
the structures of experience and the acknowledgment that
facts and values are always intertwined do not serve to
validate or invalidate specific values or ways of living. A
more sophisticated reading of Heidegger leads to this conclusion.

Bruce Foltz sees Heidegger's phenomenology as providing a new starting point for ethics, a new ethical attitude rather than a specific set of prescriptions about how
people ought to live. For Heidegger, the metaphysics of nature cannot and should not be taken as the basis for ethics;

metaphysically speaking, things in nature can only be approached as the idealized physical bodies studied by the sciences. Rather, ethics becomes possible when individuals open themselves up to the far richer and more meaningful *Umwelt* with which they are most immediately concerned. As Foltz puts it, "an ethic that is to coincide with the step back out of metaphysics must proceed from our own fundamental relationship . . . to being, that is, from the manner in which we are to hold . . . open our essence to being." This is, to borrow language from a different philosophical tradition, a meta-ethical point, which means that it is concerned with the nature and possibility of ethical judgments as such rather than with the normative principles that inform particular judgments. As a consequence, Foltz's reading of Heidegger says nothing about the content of the lifeworld or the relative importance of the connections that make it up; it does not produce a set of normative principles.

Nevertheless, Foltz does believe that this sort of thinking is useful for those involved in environmental affairs. Take the example of wilderness preservation, one of the most venerable planks in the environmentalist platform. There are two possible ways to think about efforts at preservation. On the one hand, one can remain within the modern worldview, wherein "wilderness areas can be set aside as inventories for specific sorts of recreation, which in turn allow for a more efficient output of job energies." On the other hand, one can step outside of that worldview, which opens the possibility that "wilderness areas may be genuinely saved as those places on earth where the mystery of self-seclusion consorts in splendor with the wonder of self-emergence." Foltz happens to believe that this latter view is more desirable, and hence arises the need for Heidegger's critique of modernism: it is the only way to truly step outside that which is to be criticized. As he puts it, the "uncanny dominance of technology" will "co-opt and incorporate all attempts to stand outside the technological framework" unless those attempts are "derived from, and solidly rooted in, a thinking that approaches the earth poetically." Hence, "everything depends on whether the saving arises from dwelling, and thus whether it is founded on the poetic."[25]

This is not particularly objectionable, but it should be noted that so far nothing has been established about whether wilderness should actually be preserved, or how much of it, or in what condition. There are good reasons to doubt that Foltz's insight into Heideggerian meta-ethics can be brokered into a fully fledged normative theory. To adopt Heidegger's language, authenticity is an openness to the world and a recognition that humans are confronted by choices about how they are to live in the world. It does not seem safe to assume that if humans all began to live authentically, their choices would converge automatically on a single vision of an environmentally sound lifestyle. To make this assumption, it seems to me, is to once more smuggle presuppositions into a philosophical method that is supposed to be presuppositionless. In this case, the contraband is not scientific knowledge but a specific ethical or political agenda. It is a commonplace in the environmental movement that the modern, urbanized, bureaucratized, technology-intensive, artificial lifestyle has alienated humans from nature and that a more "authentic" life would involve closer contact with the land. A certain element of nostalgia permeates this aspect of environmental thought: the vision of the authentic life is often modeled after the supposedly gentler and more harmonious life of the subsistence farmer or the pre-agricultural hunter-gatherer. To appeal to phenomenology in support of this "commonplace" without inquiring into how it came to be accepted as a commonplace is to produce, if I may say so, an inauthentic sort of authenticity.

ADEQUACY

Aside from the problem of facts and values, there is a second reason to doubt that environmental philosophers can find a moral compass. This concerns the standards by which the ethical principles or intuitions of environmental philosophers are themselves to be judged. In assessing each others' work, the issue for many speculative environmentalists is not, ultimately, whether the ecological view of the world is true or whether the set of normative claims it fosters is valid, but only whether the worldview and its attendant values are adequate to the task of supporting environmentalism as a political movement. The adequacy criterion demands only that worldviews and moral principles provide intellectual firepower to be used in defense of nature; the righteousness of the environmentalist cause and the existence of natural value are both taken for granted.

The case I have been building is that many environmental philosophers work backward from a desired social and political outcome to the first principles that can support that outcome. The evidence for this includes the tendency of environmental thinkers to carry out what I have called selective philosophy and selective (or even preemptive) science. It is in the domain of ethics, where environmental philosophers most often appeal to the adequacy criterion, that the backwardness of speculative environmentalism shows itself most clearly.

Recall that ecofeminists reject deep ecology not because its first principles are poorly defended, and not because those principles are applied inconsistently, but because the oceanic holism to which deep ecologists appeal is supposed to have political consequences that are unsavory from a feminist viewpoint. This is also the case when Callicott rejects the ecofeminist approach to ethics on what can only be called procedural grounds. By turning away from ethical theory and from the commitment to science and rationality implied in ethical theory building, Callicott argues, ecofeminists have turned away from the commitment to seek agreement through persuasion. Without such an agreement, he argues, ecofeminists can only fall back on what Cheney calls "negotiation." Callicott is concerned that such negotiation will necessarily favor those who come to the table with greater power—in this case, the despoilers of nature.[26]

Robert Elliot follows the same path when he develops his indexical theory of intrinsic value. His primary concern is not the rightness of his theory but whether it can accommodate "the greenest of green environmental values," which presupposes a standard of "greenness" that is neither stated nor defended. He is also selective in his definition of naturalness, the value-adding property on which his argument against restoration ecology hinges. He sets aside the view that everything is natural, that all that happens in the universe is determined by the laws of nature, simply because it does not support injunctions against pollution and other kinds of environmental change. He insists that "the appeal to the natural, if it is to do anything like the normative work required of it, signals a notion of the natural somewhat more fine-grained."[27]

I would distinguish two kinds or degrees of adequacy, which for convenience I label "strong" and "weak." A worldview or system of ethics is strongly adequate if it offers compelling arguments on behalf of the values and obligations of which environmentalists approve, arguments that are likely to convince dissenters and skeptics. A worldview or system of ethics is weakly adequate if it is merely compatible with the desired social or political outcome, which is to say that there is no explicit contradiction between principle and practice. In either form, the criterion of adequacy has a pervasive influence on environmental thought and serves as a kind of ideological litmus test whereby environmental philosophers are able to distinguish friends and foes.

More often than not, any way of thinking that is regarded as inadequate in either sense is rejected out of hand, regardless of its other merits. Anthropocentrism is generally regarded as a hopeless case. Many would argue that human-centered ethical theories are not strongly adequate, since they do not unambiguously forbid actions of which environmentalists disapprove. It would follow from this that anthropocentrism must, at the very least, be treated with caution. Some would make the much stronger claim that anthropocentrism is not even weakly adequate: it stands in direct contradiction to the goals of environmentalism. This latter group is more likely to view anthropocentrism as the root of environmental evil. In any case, all that is needed to discredit any ethical theory in the eyes of most environmental philosophers is to label it as anthropocentric.

Turning this around, I have given a long list of reasons to doubt that the central case of speculative environmentalism can be strongly adequate, in the sense that it could convince those who now reject the claims of environmentalists. Consider the case of a person who, after calm reflection and based on a great deal of relevant evidence, approves of something like tidiness as a marker of the degree of human control over the contingencies of the surrounding world. Imagine that this person believes not only that intrinsic value increases with tidiness but also that tidiness always trumps naturalness in calculations of intrinsic value. A great deal of ink could be spilled constructing perfectly adequate arguments for the supremacy of tidiness on this or that meta-ethical foundation. On this basis, the person may find that the local shopping mall, or a manicured lawn, exhibits that value-adding property to a greater degree than any forest; this might even lead the person to rejoice when a particularly sloppy patch of old-growth forest is cut down to construct a shopping mall that is even tidier than the shopping mall just up the road. Those who are already convinced of the rightness of technological progress might well be enthralled by such a display of reasoning, but it is unlikely that environmentalists would be swayed. Who, then, would be swayed by the arguments of environmental philosophers who used the same tactics? Environmentalists and their opponents, if such a distinction has any meaning, would be at an impasse; each group would simply assert and re-assert its own preferences, hoping against hope to convert a few of its opponents.

The truth of the matter is that many of the arguments of environmental philosophers are really directed to other environmental philosophers. At times, academic environmental philosophy seems almost like a contest in which each camp strives to demonstrate that its particular configuration of principles is the most compatible with environmentalist goals while the principles of rival camps rule out this or that desired objective. When environmental philosophers judge the results of this contest, some of them seem to confuse weak adequacy with strong adequacy, as if the mere compatibility of first principles with environmentalist values is enough to establish the literal and compelling truth of both. To take just one example, Elliot claims that he wants to convince dissenters, perhaps even the "despoilers of nature" themselves, of the intrinsic value of naturalness. What he offers instead is only the absence of a contradiction between his preferred meta-ethic and his preferred normative claims. Like many of his colleagues, he is trying to convert the multitudes by preaching to the choir.

Part 3

———

Hope

[5]

Environmentalism without Illusions

There is no need to rehearse the litany of environmental problems that now confront human civilization. From local habitat loss to global climate change, these problems have entered the public imagination through the media and the tireless efforts of environmental advocacy groups. Each problem can be considered on its own and, as such, is troublesome enough. For many, though, each problem is only one component of a much broader and more perilous environmental crisis. In the face of such a crisis, the decisions humans make over the next few years may well determine whether civilization—or the human species itself—will survive the new century.

On the largest scale, an environmental crisis cannot even be perceived as such without a great deal of scientific and technological sophistication. This implies that, given the limits of the sciences, there will always be a degree of uncertainty about the true nature and severity of the crisis. While there is some scientific consensus about global climate change, for example, disagreements remain about precisely how much the climate will warm and the consequences of that warming for local ecosystems. The scale of the crisis also implies that the crisis itself is not at all obvious. Global climate is a scientific abstraction and, as such, is removed from perception and from the concerns of ordinary life; a change in global climate is a process so vast, so complex, and so far-reaching in its consequences as to defy imagination. Little wonder, then, that climate researchers and environmentalists alike have had so much trouble in motivating people to do something about it.

In a very different sense, the environmental crisis is more immediate and much more pressing. My earlier discussion of phenomenology yielded a way of understanding the environment not as an abstract assemblage of systems but as the surrounding world of experience with which we are most concerned in our ordinary lives. A crisis within that field can hardly escape notice. "Why do I need a centralized computer system to alert me to environmental crises?" asks Wendell Berry. "That I live every hour of every day in an environmental crisis I know from all my senses." In much the same spirit, Anthony Weston has proposed that at the deepest level, the environmental crisis consists of a disconnection from the vividness and vibrancy of life in the world, a disconnection that is transforming the world "into the desolate and utterly humanized wasteland that too many of us inhabit."[1]

I do not intend to choose between these two ways of conceiving of the environmental crisis, nor will I affirm or deny the reality of such a crisis. Later, I will have more to say about the nature of environmental problems and about the intertwining of their perceptual and scientific aspects. For the moment, my point is simply that many believe humanity to be embroiled in an environmental crisis and that the crisis can be conceived in various ways and across various scales. If this belief is correct, the question remains: What grounds do we have for hope?

Speculative environmental philosophy is itself an expression of hope that the way out of the crisis is not hard to find. If the root causes are intellectual, if humanity has fallen from grace by thinking in a manner that is out of step with the natural order, then the way back into grace is also intellectual. The way to change people's behavior is to get them to think differently about their place in the natural world. The hope is that a truly adequate ecological worldview will call people to live a better kind of life in a voice that cannot be ignored. More than this, it will provide a firm theoretical foundation for making wise practical decisions, decisions that will slow or even stop environmental destruction.

Quite honestly, I would love to think that philosophy, my own chosen profession, holds the key to solving the most serious problems now facing civilization. I do not think this is so, however. People who live by the mind are prone to see everything as an intellectual puzzle. For all their well-founded skepticism about the presuppositions of modern life, environmental philosophers too can entertain false hopes about their own status and importance in the world. Intellectual honesty, guided by the principle of parity, demands that critical philosophical reflection consider the limits of its own power.

The best hope for environmental thinkers, of course, is to find and apply the correct theory, the one most firmly grounded in reality. Nothing persuades quite like the truth. However, this seems a faint hope indeed. If the Western tradition of intellectual inquiry has a legacy, it is in the successive demolition of illusions. The belief that the human mind can establish certain knowledge, a kind of easy familiarity with the cosmos, has been dashed time and again. What remains in the realm of the sciences is a set of more or less powerful models that are nonetheless tentative, restricted in scope, and viewed only against a backdrop of profound and intractable uncertainty. This uncertainty places limits on the conclusions to be drawn from the sciences, especially ecology: not only are the living interactions studied by ecology complex but they exhibit neither a single state of equilibrium nor any sort of beneficence. As a consequence, current ecological theory does not promise a return to primal innocence or even some sort of natural moral standard against which to measure human activities.

THE SOLUTION OF OUR PROBLEMS

Perhaps a better way to get at the limits of hope is to reconsider the nature of environmental problems themselves. George Lakoff and Mark Johnson tell the story of an Iranian student at Berkeley who heard what he thought to be a beautiful metaphor embodied in the expression, "the solution of my problems." He envisioned

> a large volume of liquid, bubbling and smoking, containing all of your problems, either dissolved or in the form of precipitates, with catalysts constantly dissolving some problems (for the time being) and precipitating out others. He was terribly disillusioned to find that the residents of Berkeley had no such chemical metaphor in mind.[2]

Instead, they had in mind a puzzle metaphor, according to which any given problem has a single, correct solution; once the problem is solved, it is solved forever.

Speculative environmental philosophy has been governed by the puzzle metaphor, offering an ecological worldview as the correct solution to the perceived crisis. I would suggest, however, that the chemical metaphor is a better fit with human experience, especially where environmental problems are concerned. A shift to the chemical metaphor leads to some striking observations regarding both the nature of environmental problems and the prospect for their (dis)solution.

The first observation is that the environmental crisis is not a single monolithic puzzle awaiting a solution. Indeed, contrary to the radical tendency of environmental thought, the crisis is not rooted in a particular form of society or in a particular way of thinking. Instead, the crisis is an ever-present (if sometimes latent) potential within the human condition; it is an unavoidable and perhaps even indispensable part of human life on Earth. Environmental problems are, in a word, endemic.

The reason for this pervasive potential is that humans are neither wholly natural nor wholly free. Instead, the human condition seems to consist in being trapped between these two poles. On the one hand, humans are living organisms and, as such, we are vulnerable to and dependent upon our surrounding environment. On the other hand, moral freedom imposes on us the necessity of choosing and acting on the basis of nothing more than our own limited understanding of ourselves and of the surrounding environment, motivated by our own interests and values; there is no way to guarantee that our choices and actions are wise. Because of our vulnerability and uncertainty, we humans are perpetually getting ourselves into trouble of one kind or another.

To continue with the chemical metaphor, the vulnerability and uncertainty of the human condition mix together in a solution. The choices and actions of individuals and groups of humans occasionally precipitate particular problems out of this solution. This is a bit simplistic; catalysts include not only choices and actions, as though they could exist in isolation, but also institutions, modes of production, patterns of settlement, and so on, not all of which are under direct human control. Global climate change, for example, is the product of a very specific configuration of technological, economic, social, political, and ecological factors, a configuration that seems almost to have its own internal logic and direction.

Also, the problems brought out by a given pattern of human life may be more or less severe. Clearly, modern industrial capitalism has much to answer for. It is reasonable to suppose that this particular complex of technological, social, economic, and political systems has precipitated out more—and more serious—problems than any other complex in human history. Just as clearly, no group of humans is entirely "innocent" of environmental problems. A modern developer laying out a suburban subdivision and a pre-Columbian Native American setting a prairie fire to flush out game are both altering their environment; both are setting in motion a chain of consequences that cannot be predicted with certainty, and both are vulnerable to those consequences. The culture of one might be commended for the relatively modest scope of the alterations it brings about, but neither culture exists in a state of primal innocence or harmony with nature. Because environmental problems are endemic, there is no way of knowing what innocence or harmony would look like.

Another consequence of the chemical metaphor concerns the manner in which problems are to be addressed. To live by the chemical metaphor, Lakoff and Johnson note, would be to accept it as a fact that no problem ever disappears forever. Rather than direct your energies toward solving your problems once and for all, you would direct your energies toward finding out what catalysts will dissolve your most pressing problems for the longest time without precipitating out worse ones. The reappearance of a problem is viewed as a natural occurrence rather than a failure on your part to find "the right way to solve it."[3]

It follows from this that environmental philosophers, and anyone else concerned with environmental problems, should have modest expectations for their own work.

As with all metaphors, this way of regarding environmental problems has its limitations. The chemical metaphor implies, for one thing, that problems can always be recognized as such, and that there will be widespread agreement that they exist. A precipitate, after all, is a solid: it has tangible substance to it that cannot be denied. This is clearly not the case with many environmental problems, the very existence of which can be a matter of debate. Someone might regard global climate change not as a problem but as an opportunity for, say, trans-polar navigation or year-round gardening. More seriously, those who depend on exploitation of forest resources for their survival may not care or even notice that their activities bring about the extinction of several species of insect; the problem only appears as such to some groups of environmentalists and biologists. Even if a problem is widely recognized and the effort is made to dissolve it, there may be disagreement over the degree to which those efforts are successful and when they should be brought to an end.

CONFLICT
The limitations of the chemical metaphor suggest a somewhat different approach to the matter. Perhaps environ-

mental problems are best understood as conflicts among humans or groups of humans. It is a simple matter of observation that there is seldom widespread agreement among humans about which goals are worth pursuing or what might be the consequence of pursuing a particular goal. One group makes a decision and acts on it, bringing about some change in the surrounding environment. That change may threaten, limit, or destroy the opportunity for another group to pursue one or more of its goals. Without a sure moral compass, there is no basis for claiming that something has gone wrong with the environment in terms of some universal standard of rightness; there is simply a conflict. More basically, environmental problems understood as conflicts do not have an existence of their own that is separate from the perceptions, goals, and values of those who are engaged in the conflict.

A number of variants of environmental conflict should be distinguished. The first of these is direct conflict between people. On the small scale, consider the case of my neighbor's dog, which might fall into a barking fit at any time of day or night. My interest in peace and quiet—not to mention a good night's sleep—is pitted against the value my neighbor finds in dog ownership and the presumed negative value of letting the dog sleep indoors at night. It could be argued that the best way to resolve the problem—the one that is most likely to be effective without entailing even worse consequences—would be through some sort of reasoned negotiation with my neighbor.

More serious is the complex relationship between a factory that generates pollution and the community within which it operates. On the surface, the main values involved are the corporation's pursuit of profit for its shareholders, on the one hand, and the community's interest in maintaining the health and well-being of each citizen on the other. Other values must also be taken into account, however, including the benefits to the community of employment and tax revenues provided by the corporation and the corporation's interest in maintaining a "good neighbor" image, among others. Solving the problem will be a matter of deciding, as a community, which goals should have priority over which others and seeking the appropriate balance in practice.

The second variant involves conflict among an individual's own values, where the pursuit of one undermines the pursuit of others. All too often, the best-laid plans of even the most thoughtful people fall prey to the law of unintended consequences. Examples are abundant and do not require a great deal of elaboration: the comfort of air conditioning and the convenience of aerosol sprays have long since been linked with increased exposure to ultraviolet radiation and hence to greater risk of skin cancer; the availability of plentiful and cheap fossil fuels has been linked to local air pollution and to global climate change. In the domain of environmental conflict, the natural sciences can help to illuminate such unpleasant surprises, even if they cannot eliminate them.

A third variety of conflict nestles between these two, in which the pursuit of a given goal undermines the possibility of others pursuing them or of anyone pursuing them in the future. The use of finite and non-renewable resources is self-defeating in this way. The consumption of fossil fuels is an obvious example. There is no doubt that an economy based on abundant cheap fossil fuels has its advantages, at least in the short term; it is increasingly difficult for many in the industrialized world to imagine life without them. But we must imagine it; because we are using up fossil fuels today, our grandchildren will not have access to them. Whatever the other unintended consequences may be, industrial societies are only now beginning to face up to the fact that a fossil-fuel economy is not sustainable in the long term.

Environmental disputes frequently involve some complex mixture of these three kinds of conflict. Take, for example, recreational enjoyment of what are usually thought of as wilderness areas. This case is particularly revealing, because it has built into it some of the complexity of what people value about "wild nature." For some few, the enjoyment of wilderness is founded in the intellectual and aesthetic appreciation of ecosystems in a state that is relatively unaltered by human activities. For others, its value lies in the potential for contact with interesting plants and animals. Still others enjoy attractive landscapes, court danger, crave solitude, and so on. Wilderness is difficult to define, in part because so many areas associated with the ideal of wilderness are managed to a greater or lesser degree in order to maintain some set of attributes that is of value to people—but which attributes are most important? To the extent that the value of wilderness is in dispute, the definition and the character of wilderness areas will be disputed as well.

The value of danger and the value of peace and quiet frequently come into direct conflict. Someone's enjoyment of a lake for water skiing can spoil any hopes I have of observing water birds or swimming or simply enjoying the tranquility of a sunset reflected in the water. The same applies to the value of access. Widespread development of the lake might destroy all that I value in it, but protecting the lake from all development might thwart my hopes of building a cabin there to ensure easy access to peace and quiet.

Ironically, it often happens that the more widely a value is shared and the more it is pursued, the harder it is to find. People have been visiting U.S. national parks in record numbers; traffic jams are not uncommon in Yellowstone. This has been aggravated by the combination of the value of wilderness with the value of comfort and security as many try to "rough it" in large, air-conditioned recreational vehicles with microwave ovens and satellite television.[4] A similar pattern has governed the suburbanization of America as people have moved out of the city to be closer to what they imagine to be the unspoiled countryside. A complex array of values, including those placed on security and mobility, has motivated people to bring much of city life with them. As a result, many of the values sought in the suburbs are no longer to be found there, and much that was of value in small towns and cities has been lost as well.

Conflicts about pollution are at least as complex as those about wild nature. The idea of pollution has strong moral overtones; it suggests that whatever has been polluted has been rendered impure and that all such impurity is inherently evil. The first point to acknowledge about pollution, from a skeptical point of view, is that the very idea is relative, leaving little room for sweeping condemnation. Human activities—indeed, the activities of all living organisms—tend to result in the movement of material from one area to another. Sometimes the results are desirable, as when water is brought into my house through pipes or grain is sent by rail from Kansas to Chicago. Sometimes the results are undesirable, as when sewage from my house is introduced into a nearby river or exhaust from the diesel engines that haul the grain changes the composition of the nearby air. Any product of human activity changes the natural system into which it is introduced in ways that are difficult to predict.

The question to ask about the dislocation of material

is not whether the outcome is good or bad in any absolute sense, but simply whether and to what extent it impairs the ability of others to pursue their various goals. Pollution, then, can be defined as the introduction of material into an environment that degrades characteristics of that environment that are of value to some group of people. Of course, some cases of pollution meet with nearly universal condemnation, such as the fallout from Chernobyl that degraded or destroyed that which nearly everyone values: human health and well-being itself.

Most cases of conflict over pollution are really very complex. On the largest scale, the combustion of fossil fuels has been bringing about a gradual change in the chemical composition of the Earth's atmosphere. This is not necessarily an unmitigated evil; there are likely to be winners and losers in the wake of any resulting change in the Earth's climate. I will grant that, given the severity of some climate change scenarios, the losers are likely to be in the majority. Simply consider that the densest human populations are to be found within a few miles of the ocean: a rise in sea level of even a meter will threaten some of the world's largest cities. Even so, the point remains that a change in the climate is not bad simply because it is a change or simply because it has been brought about by human activities. Rather, global climate change, like any instance of pollution, must be judged on the basis of its impact on a wide array of human values. For most people, global climate change will probably be a very bad thing indeed.

Efforts to identify and ameliorate particular instances of environmental damage are also subject to conflict, which is aggravated by the complexity of environmental systems and the uncertainties of the natural sciences that study them. It is often difficult to prove that there is a causal connection between the introduction of new materials into a given system and particular instances of harm. This has long been the case with debates over the effects of certain synthetic chemicals, for example; claims and counterclaims are launched by opposing camps of scientists and advocates. The best that a skeptic can do, it seems, is to pick those studies that seem the least disingenuous, filtering out the undue influence of industry and of environmental groups alike. Even this filtering is unlikely to produce a clear picture.

The problem with predicting the course of global climate change, for example, is that the global climate is too complex to be adequately modeled. The changes that have been observed over the last few years might just as easily be the result of previously existing cycles in global mean temperature in an interglacial period. Further, despite the temptation to the contrary, there is little basis for saying that the severity of a particular storm or of a particular climate phenomenon, such as the 1997–1998 El Niño, can be attributed to increased levels of greenhouse gases in the atmosphere. The point here is simply that the sciences must play some part in addressing environmental problems. But the sciences always, necessarily, leave room for doubt.

This account of environmental conflicts is far from complete. A number of unanswered questions remain about, among other things, the range of values and goals that may be involved and the processes by which particular disputes might be settled. Beneath these questions lies the still more basic question of whether it is accurate or advisable to understand environmental problems entirely in terms of conflict. At the very least, I believe there is a good deal to be learned by understanding problems in this way, especially since many other environmental thinkers adopt this same understanding, if only tacitly. Assuming my characterization is accurate, what do environmental philosophers see themselves as bringing to ongoing environmental conflicts? What hope do they offer?

PHILOSOPHY AS ADVOCACY

Most of my argument to this point has been based on the assumption that environmental thinkers take their own speculations seriously as the literal truth. Many of them might agree with me that there is no possibility of attaining absolute certainty about the world and the place of humans within it but insist that uncertainty is beside the point. Whatever their rhetoric, some environmental philosophers might admit that their underlying reason for doing environmental philosophy is not to arrive at the truth, perhaps not even to provide a single correct solution to the environmental crisis. Instead, it is to defend a perspective or a set of values that has not received the attention it deserves in efforts to resolve (or even recognize) environmental conflicts. If this is so, then the hope of environmental philosophy is simply that the value of the natural, the wild, the particular, or the local will be accepted as a legitimate basis for decisions in the public realm. Even better, some of the opposition, or the public at large, might be persuaded that such value should overrule decisions made purely on the basis of economic of political expediency.

In my discussion of the adequacy criterion, I charged environmental thinkers who construct whatever system will serve their political aims and then insist on its truth with disingenuousness. At least some would shrug off this charge, embracing backwardness as a virtue and explicitly using philosophical principles as a means to social and political ends. It is fairly clear from his writings, for example, that Aldo Leopold developed his land ethic because it sup-ported the goals of conservation. Because the sciences do not provide certainty about how biotic communities work and because economic incentives are not enough to foster conservation, the development of a land ethic became a social necessity. At no point did Leopold advance the claim that his ethic was grounded in truth.[5]

Although many deep ecologists seem to be true be-lievers, Arne Naess is more pragmatic. He allows for a range of possible worldviews under the umbrella of deep ecology, calling his own configuration of principles "Eco-sophy T" to distinguish it from other legitimate possibili-ties. According to Naess, it does not matter what principles a deep ecologist adopts, as long as they are identifiably "ecological" and as long as they are opposed to shallow environmentalism.[6]

More recently, there has been movement along sev-eral fronts toward a more explicit recognition of environ-mental philosophy as advocacy. Postmodern environ-mental ethics and environmental pragmatism are both interesting, from a skeptical point of view, in that they step away from the speculative mainstream of environ-mental thought. Each carries with it the risk, however, of slipping into a philosophical discourse constructed almost entirely ad hoc in the name of environmental advocacy.

Postmodern environmental ethics is predicated on the principle that there is no single privileged discourse, no way of speaking or thinking that can claim to have access to the way things really are. Truth, in this account, is so-cially constructed through a process of negotiation that takes place in a particular social context. Jim Cheney, one of the first to develop a postmodern environmental ethic, proposes a negotiation about bioregions, "a complicated working out of the relationship between home, identity, and community" which will, he hopes, provide the value orientation that is needed for the resolution of environmental problems.

Not all participants in the negotiation are equal in Cheney's eyes, however. He holds that a discourse may be (legitimately) privileged insofar as it makes it possible to see "distortions, mystifications, and colonizing and totalizing tendencies within other discourses." So the currently dominant paradigm of modern science is automatically discounted, as are its technological, political, and economic adjuncts. At the same time, some form of ecological paradigm attains privileged status simply because it undermines the privileged status of the scientific paradigm. Cheney implies that any discourse that stands in opposition to the dominant paradigm without itself falling into some equally distorted metaphysical view is "true" in this sense.[7] Widespread acceptance of this conclusion would clearly put environmentalists in a politically advantageous position.

As for environmental pragmatism, Andrew Light and Eric Katz emphasize that it is not simply "a new side to the environmental ethics debate" grounded in the "established sources" of philosophical pragmatism, such as the writings of John Dewey or William James. Rather, they see it as an attempt to raise new and important questions in environmental philosophy that are "inspired by the philosophical legacy of pragmatism." In particular, they would foster a "metatheoretical pluralism" that argues that environmental philosophy is contested terrain. Like Cheney's call for negotiation, metatheoretical pluralism is an interesting proposal that deserves attention.

For Light and Katz, however, there seem to be practical and political limits on how far afield environmental philosophers may go. The terrain of environmental philosophy is contested, they argue, but only with regard to the means that may be used to achieve the existing policy objectives of the environmental movement. The contest will yield "a variety of theories to account for, encourage and justify values in nature, and to explain and validate environmental policies." Notice that the goal is not to discover whether there are values in nature or to rethink environmental policies. Light and Katz seem to assume that environmentalists already agree on the correct set of values and policies; all that remains is to "explain and validate" them. Bryan Norton makes this explicit in his "convergence hypothesis," according to which environmentalists tend to agree on policy objectives even if they disagree on worldviews and value frameworks. The unity of environmentalism is to be seen not in its rhetoric—"the vocabularies in which environmentalists argue their political case" —but in its cohesiveness as "a force in public policy."[8]

On the surface, environmental pragmatism seems to offer only a cynical kind of hope: if philosophers do not have to be too picky about theory, they can do whatever it takes to persuade more people to join the environmental cause and to provide stronger arguments to use in the public arena. This superficial reading is not really fair to Light, Katz, and Norton; I will do what I can later to make it up to them. For now, though, the cynical version of environmental pragmatism serves to highlight the practical limits of environmental philosophy more generally. I have argued that many speculative environmentalists essentially argue backward from a pre-approved social and political agenda to whatever theoretical edifice seems to offer support. By comparison, the cynical version of environmental pragmatism at least has the advantage of being honest; it openly embraces the backwardness of environmental thought, making a philosophical virtue of political necessity. Even on purely pragmatic grounds, however, such honesty does not overcome two basic practical problems of ad hoc reasoning.

The first problem is that arguments formulated backward are likely to convince only those who already agree with the conclusions. Writing in 1993 as the editor of *Environmental Ethics,* Eugene Hargrove worried that the field of environmental ethics did not seem to be having much of an impact. Solving this problem, he urged, "should be the top priority of everyone concerned about the environmental crisis." At least part of the problem, he argued, could be attributed to a lack of interest in philosophy on the part of the general public. I would argue that the problem lies much closer to home.

It is no doubt true that the erudite arguments of academic environmental philosophers may not be convincing or even accessible to the general public, even if they are fully adequate by the standards of other academics. It is not all that difficult to understand how people who are simply trying to live their lives could fail to be moved by complex metaphysical and moral reasoning about, for example, the intrinsic value of trees. Even so, there is no reason to believe that a more philosophically sophisticated public would necessarily be any more likely to side with Hargrove and his compatriots. If it is true that academic environmental philosophers have, by and large, been preaching to the choir, then Hargrove might as well say that the reason environmental ethics is not having an impact is that there just are not enough environmentalists out there. In a follow-up report written five years later, Hargrove conceded that theoretical work in environmental philosophy was at last having an impact—mainly with such groups as the Society for Conservation Biology and the Society for Ecological Restoration, which are hardly comprised of hard-core dissenters.[9]

The second problem with constructing arguments backward is more urgent. Suppose that all attention is focused on providing arguments for a given set of policy objectives and that the objectives themselves are simply taken for granted. If they are not subjected to constant critical scrutiny from within, the policies put forward by environmentalists might harbor all sorts of practical dangers and difficulties. In short, widespread adoption of the kind of ecological worldview put forward by many environmental thinkers might just as easily aggravate environmental problems as solve them.

The managers of Kenya's Tsavo National Park, for example, assumed that the best management policy for the elephant herds in the park was no management at all, on the principle that human interference could only bring about an imbalance in any natural system. They hoped that, left alone, the elephant population would find its own natural equilibrium. The result of this policy was a disaster; the elephant population exploded then collapsed during a severe drought, taking much of the vegetation of Tsavo with it.[10] This by itself is not an indictment of environmental philosophy, even if the dictum that "nature knows best" is a common feature of ecological worldviews. However, there is nothing in the arguments of environmental philosophers per se that can foresee or forestall these kinds of practical consequences.

The phenomenological argument provides a further illustration. I have already established that the argument from phenomenology often boils down to a case of selective philosophy with bits of scientific terminology smuggled in for good measure. The Heideggerian form of this concerns the "authenticity" of an individual human life, although there are two separate ways of understanding the moral force of that term. When someone takes authenticity to be a specific standard against which human practices should be measured and then experiences a return to agrarian or pre-agrarian subsistence as authentic, their judgment cannot be gainsaid on the basis of further application of the phenomenological method. If it should turn out that the widespread adoption of one of these visions is environmentally disastrous—as vast areas come under increasing pressure from well-meaning foragers and farmers, for example, as forests vanish and the air fills with the smoke from countless wood fires—there is nothing in a phenomenological analysis of being-in-the-world that could prevent this. At some point, the phenomenological method as such ceases to be of any help. Criticism or correction of an "authentic" or "harmonious" vision can only be made on some other grounds, whether they be scientific, economic, social, or political.

It might be argued that once these practical consequences are known, environmental philosophers can take them into account ad hoc as they formulate their next set of arguments. So an advocate of the phenomenological approach might contrive an argument to the effect that some sorts of pre-agrarian subsistence are more appropriate than others. This does little to deflect the charge of arbitrariness or to bolster the persuasive force of environmental philosophy. More to the point, it offers scant basis for confidence that environmental philosophers can foresee the next set of unfortunate side effects. When arguing backward, it is only too easy to foster illusions and false hopes simply because those illusions seem, superficially, to support the values and practices of which environmentalists generally approve.

At this point, environmental philosophers do not seem able to offer much hope at all. It is unlikely that they will seize upon the one correct (intellectual) solution to the environmental crisis. Even if the crisis is understood in terms of conflict, it is not much more likely that environmental philosophers will be able to convert enough people to their side to turn the political tide in their favor. Finally, even if the tide did turn, there is no guarantee that the consequences would be to environmentalists' liking. There seems to be only one avenue left open. Rather than engage in direct advocacy for one side or another, environmental philosophers might focus instead on the processes by which decisions are made. For at least some environmental philosophers, the last best hope resides in pluralist, participatory, democratic processes in which a wide range of values, visions, and voices might be heard and taken into account.

THE DEMOCRATIC IDEAL

The ideal of participatory democracy is attractive and simple enough in principle. People of good conscience should be able to come together to discuss their differences openly and without fear. If the process is well designed, the groups involved should even be able to transcend their apparent differences and forge a sense of common purpose. If they are able to arrive at this point, the participants should then be able to reach decisions that are both fair and practicable. If they are particularly creative, the participants might even devise a "win-win" solution (to use the language of negotiation) in which everyone gets more of what they want than they would have gotten had they resorted to other means.[11]

In practice, of course, participatory democracy is both complex and difficult. Not all people are of good conscience, and not all people can speak without fear. Decision-making processes themselves are often poorly designed, excluding some, giving too much power to others, and squelching creativity across the board. I do not mean to imply that the ideal should be abandoned. On the contrary, I think the goal of reasonable, fair, and practicable decisions reached by democratic means is worth pursuing, even if the pursuit is difficult. I do mean to imply that those engaged in the pursuit should not be naive about its difficulty. I would even venture to say that, at least from a skeptical point of view, hope might reside precisely in not being naive: it is only when we see clearly the factors that complicate environmental conflicts that we can begin to improve the processes by which those conflicts are mitigated. As a start, I will consider a handful of complicating factors, many of them suggested by environmental thinkers who have already begun to work along these lines.

First, a wide range of values and perceptions may be at stake in a conflict. There is a strong tendency to see environmental conflict in essentially economic terms: the economic interests of stakeholders conflict, and the resolution of the conflict must lie in "balancing" those interests through a series of tradeoffs. Most of the examples I considered in my brief taxonomy of environmental problems had this shortcoming, focusing mostly on material interests and seemingly arbitrary preferences. And yet humans pursue all sorts of goals, informed by all sorts of ways of perceiving the world: self-development, aesthetic enjoyment, recreation, understanding of complex systems, adrenaline rushes, aimless reverie, ego-stroking, sympathetic connection with non-human creatures, and so on to infinity. It is the complexity and variety of human visions of the good life that make environmental conflicts so daunting—and so compelling an object of study.

Anthony Weston argues that the environmental crisis consists of more than threats to the "fragile, battered, agglomerated, homogenized 'environment' featured in ordinary environmental rhetoric." Because that rhetoric can only refer to the environment in the abstract, because it is confined to a "human, urban, economic and political perspective," it is necessarily disconnected from the immediacy, the detail, the richness, and the wildness of life in the world. In response, Weston proposes a different kind of environmentalism, the goal of which is not to attain "some sort of privileged metaphysical profundity" but to learn and then inhabit "the infinity of stories and possibilities and connections" that open up for human life in the world when we are paying attention. "How shall we now live?" he asks. "There is no single answer. There are only a multiplicity of possibilities, springing up everywhere, like a genuine revolution, like weeds in a garden."[12] One implication of Weston's approach is that, at its best, environmental philosophy can serve as a way of exploring the full range of values that are at stake in environmental conflict, values that are embedded in richer and more complex ways of perceiving and living in the world.

This may be the best point at which to respond to a likely objection to the entire project of skeptical environmentalism. Many will be ready to dismiss my argument as yet another apology for anthropocentrism. There is a kernel of truth to this, in that I am considering only conflicts that arise among human beings. I believe I am justified in doing this, given the doubts I have raised about efforts to develop a non-anthropocentric ethic. Quite simply, it seems unlikely that we humans can find any other basis for making decisions than our own values, goals, and interests.

The charge of anthropocentrism is generally supposed to imply some measure of guilt. In its strongest form, anthropocentrism is much more than just human self-centeredness: it implies that the natural order itself is human centered and exists solely to serve our interests. Many environmentalists argue that humans have brought the environmental crisis crashing down on us through this kind of blind arrogance, hence the guilt implicit in anthropocentrism. In light of what I have said about the vulnerability and uncertainty of the human condition, however, it should be clear that I would not advocate anthropocentrism in this narrowest and most pernicious sense. I do not believe that humans or human interests have any special status in the cosmos, and I do not believe that we can do as we will with impunity. I do not favor ignorance and arrogance. I do favor caution and modesty.

This is not the only form of anthropocentrism to be targeted for criticism, however. At the very least, anthropocentrism is thought to focus too narrowly on what amounts to economic interests: the fulfillment of human needs and desires. Critics argue that such a focus perpetuates environmental degradation. This is a fair criticism. It is reasonable to suppose, however, that the problem with narrow anthropocentrism is not that it is human centered but that it is narrow: it precludes many legitimate human values that lie beyond survival, security, comfort, and leisure. As Weston suggests, human life can and should be much richer than this. If I am an anthropocentrist, then, I am so only in a manner that is both modest and broad: I acknowledge that humans are not of central importance in the cosmos, and I acknowledge a range of legitimate values beyond those that may be measured in dollars.

The diversity of values and perceptions at stake in environmental conflict is compounded by diversity within the environmental movement itself: environmentalists do not speak with one voice and need not converge upon a single set of policy objectives. David Scholsberg has gone so far as to say that "there is no such thing as environmentalism"; the term is just a convenient label for a "diverse array of ideas that have grown up around the contemplation of the relationship between human beings and their surroundings." The field includes romantic preservationists, steady-state economists, planners, conservation biologists, religious evangelists, nature consumers, and a host of others. If environmentalism is often presented as a single movement with a single vanguard, it is only because most efforts to classify or analyze environmentalism have excluded or marginalized parts of the movement.[13] This suggests that there may not be a clearly delineated environmental "side" in a given dispute or even an anti-environmental side; at best, there may be a set of shifting coalitions with fluid boundaries.

The fact of diversity is not as important as the depth of diversity. If only economic interests were at stake, and if resolution could be reached by shuffling measurable quantities of natural resources from one party to another, then the success could be achieved simply by getting the arithmetic right. Conflicts run far deeper than this, however, and the parties involved might all perceive the problem and the process differently; they might even disagree about what a fair resolution would look like. To one degree or another, it is as though they are speaking different languages, even if they all use the same words. The problem of communication is further complicated when various experts come on the scene who speak the specialized languages of their own academic fields—and have motivations of their own.

The problem of communication is pervasive, but it reaches its greatest extreme in cross-cultural disputes. Not only might the parties use different words and grammatical rules, but they may share very few cultural points of reference as well. In considering such disputes, Deane Curtin wonders how the process of resolution can be "democratic" if there is no common moral language, no agreement about basic notions of fairness, rights and responsibility. He is particularly concerned about the violence that can be done to a culture when principles and practices are imposed from the outside, as when American-style solutions to environmental problems are imported to developing nations. Curtin attempts to define "democracy" in a way that does not presuppose Western, especially American, moral language; he calls it "the normative dialogical space that opens up through transformative moral experiences in which grievances can be mediated nonviolently." The key to opening up this space, Curtin holds, is to find a way to translate between moral worlds.[14]

Aside from the diversity of the participants, there is a different category of complicating factors that concern equity and access. Some groups are excluded from decision-making processes, and not all who are included are treated equally. In the United States, the environmental justice movement has arisen in response to case after case in which environmental degradation has been intertwined with racial discrimination. In general, those groups that have traditionally lacked political and economic power often bear a disproportionate share of the costs of environmental change, often without reaping the benefits. In response, activists have developed what Robert Bullard calls the "environmental justice framework," the goal of which is "to make environmental protection more democratic. More important, it brings to the surface the *ethical* and *political* questions of 'who gets what, why, and in what amount.'" Mainstream environmental organizations have also come in for criticism for what some see as "their disregard for a wide variety of environmental hazards faced by people of color, a paternalistic attitude toward low-income and minority communities," and the lack of diversity within their own organizations.[15]

Other historical patterns of exclusion and oppression also come into play. At the global level, the legacy of colonialism plays itself out in the conflict between industrialized nations ("the North") and developing nations ("the South"). Ironically, much of the bitterness of the South arises when it perceives the North as interfering in and undermining the development of the South in the name of environmental protection. It is not hard to understand, for example, why people in a developing nation would resent and resist demands that they reduce their consumption of fossil fuels. A handful of wealthy and powerful nations have benefited from the use of fossil fuels, and in so doing they have created a global environmental problem. While those nations do little to curb their own excesses, they warn of immanent catastrophe should the developing nations attempt to attain the same level of wealth and power by the same means. Meanwhile, the developing nations, who have so far contributed the least to increasing the concentration of greenhouse gases, will also be least able to cope with the consequences of global climate change—precisely because they lack the resources made available by advanced scientific and industrial development. This smacks of injustice on the grandest scale and continually frustrates efforts to reach an effective international agreement on greenhouse gas emissions.

Given all of these complications, participatory democracy is unlikely to reduce environmental conflicts. Even if decision-making processes are fair and inclusive, even if all participants engage freely in a creative and constructive dialogue, disagreements are likely to persist about even the most basic of questions: What is the meaning of human life on Earth? What can we know of our lives here? What is of value? What ought we to do? At this level, at least, environmental conflicts may be entirely intractable.

Even so, two sources of hope remain. The first is that even if conflicts run deep, a really creative decision-making process can lead to the settlement of particular disputes, at least in the short term. The conflict/dispute distinction comes from the work of dispute-resolution professionals. Environmental conflicts are "long-term divisions between groups with different beliefs about the proper relationship between human society and the natural environment." These conflicts play themselves out in "the seemingly endless series of incremental disputes concerning the enactment of specific policies." The settlement of a dispute determines a momentary balance between the positions, but the balance itself "remains a matter of continuing conflict in which a never-ending series of disputes leads to decisions which move social policy back and forth between competing positions."[16]

This seems like a grim assessment, in part because it presupposes that policy is driven entirely by competition among fixed interest groups, each of which is trying to gain power over the others. Nevertheless, when coupled with a more openly pluralist model of democratic participation, the distinction between disputes and conflicts offers some hope. Even if there is no way to resolve deeper conflicts, the effort to make decisions by democratic means can occasionally lead to settlements that acknowledge and balance a wide range of competing values. More than this, the endurance of the underlying conflict implies that democratic efforts to settle disputes can and must be constantly pursued and renewed. If they are, then there is always the possibility of reaching still more creative and workable settlements to disputes in the future.

A further consideration might help to brighten the picture. The literature of dispute resolution tends to regard human interests as fixed, at least for the duration of the dispute resolution process: the goal is to accommodate the preexisting interests of the stakeholders, not to judge or tamper with those interests. This overlooks, I think, the potential of open democratic deliberation to change its participants. Deane Curtin distinguishes two functions of what he calls ethical discourse: reiterative value reinforces the moral universe of the community, but transformative value is "subversive" because it opens up new ways of understanding. In the context of decision-making, this means that a modicum of "moral imagination" on the part of the participants might lead them beyond their own prior expectations and interests. To what end? That is difficult to say; "transformative value can never be anticipated," Curtin notes.[17] What is important is that such a transformation is a possibility for *all* participants . . . including those who identify themselves as environmentalists.

WHAT'S A PHILOSOPHER TO DO?

I have begun to make the transition from the largely negative phase of skeptical environmentalism—the critique of speculative environmental philosophy—to what

I hope will be a more constructive phase. Rather than merely harping on what is not possible, I have begun to outline a vision of what may be possible within the limits of knowledge. In closing, I should say something about the role that philosophers might play within this vision. If it is really the case, as I have argued, that environmental philosophers are better off abandoning the speculative project, what might they do instead?

Like Light and Katz, I do not intend to establish another camp within environmental philosophy as it now exists. Instead, I intend to point out a new direction for environmental philosophy, a different way of approaching the constellation of intellectual and practical problems of human life in its environmental context. In doing so, I must acknowledge that I am not alone in this and that I have learned a great deal from others who have also taken steps away from speculative environmentalism. Environmental pragmatists, postmodern environmental ethicists, those concerned with environmental justice, some ecofeminists, and some social ecologists have made significant contributions, as have a number of thinkers and writers outside of academia. The most I can do is to make a new and (I hope) significant contribution of my own to this movement.

I begin with the recognition that despite all of my concerns about environmental philosophy, philosophers have an important role to play in public discussion of environmental problems. I would even go so far as to say that some sort of philosophical reflection is a necessary condition for effectively addressing those problems. As deep ecologists and numerous others have pointed out, there can be no merely technical solution to a crisis with intellectual, moral, and spiritual dimensions. The problem lies not only with the tools we do or do not have at our disposal but with the manner in which we wield the tools we have. Writing of the problem of specialization, Aldo Leopold remarked that "there is one vocation—philosophy—which knows that all men, by what they think about and wish for, in effect wield all tools. It knows that men thus determine, by their manner of thinking and wishing, whether it is worthwhile to wield any."[18]

The ways of living (and of using tools) that make up modern civilization are all too often based on a narrow conception of what is possible and what is of value for human life, and sometimes they are based on a willful self-deception regarding the security of human tenure on this planet. The power of what I have called the radical strand of environmental thought lies precisely in its ability to expose the illusions at the core of our self-understanding. Philosophers have a good deal to contribute to this effort.

The key is for environmental philosophers to adopt a more practical stance and to rethink the relationship between theory and practice. Bryan Norton has distinguished two ways of conceiving this relationship: applied philosophy and practical philosophy. The central case and its variants are examples of applied philosophy, in which the task of the philosopher is to discover and defend the theory that will foster whatever are considered to be the appropriate practices and policies. When there is a dispute over policy, philosophers try to provide arguments that can be used by the practitioners of environmentalism to convince or persuade dissenters. When the arguments are ineffective, if important parties to the dispute remain unmoved, then the philosophers must return to the theoretical drawing board in order to formulate more adequate arguments.

Practical philosophy, on the other hand, involves a much more direct engagement in efforts to address environmental problems. From a skeptical point of view, the most important implication of this is that philosophers should abandon the tidy certainty of speculation. According to Norton, the practical philosopher should get involved with the concrete details of environmental problems in all their complexity and gain a clear understanding of the full range of values that are in dispute; only then should he or she move toward more general principles in the process of building consensus on policy and practices.[19]

Environmental pragmatism carries with it the danger of slipping into an "anything-goes" kind of advocacy: any theory will do, as long as it is green enough. There is a simple way to avoid this problem. To the extent that they take up the practical stance, environmental philosophers should be more daring than they have been up to now. This may seem ironic, given that I have been urging caution and modesty, but my skeptical assessment of their work has led to the conclusion that environmental philosophers are often too conservative: they hold certain values and objectives in reserve, beyond the reach of critical scrutiny. Instead, I would urge environmental philosophers to engage in a philosophical discourse in which everything is at stake. Rather than contenting themselves to gather together theoretical means to a predetermined practical or political end, they should call those ends themselves more sharply into question. Norton is clear on this point. He does not hold the current aims and policies of the environmental movement to be sacrosanct; in his version of environmental pragmatism, even core values may be questioned and made subject to revision in the course of public debate.[20] Herein lies the possibility of transformation.

In taking up the practical stance, environmental philosophers need not abandon all concern for theory. There remains a good deal of theoretical work to be done, even if it is focused upon and directed by practical concerns. For example, I have made a number of claims about the human condition that call for further investigation. One theoretical project along these lines might be to consider the intertwining of freedom and nature more closely, drawing from the philosophical and scientific traditions. Continental philosophy since Kant has had a great deal to say about human freedom, its scope, and its limits, while recent scientific developments have apparently begun to undermine the notion of free will. Rather than selecting only those perspectives that appear favorable, environmental philosophers should weigh these traditions carefully and critically in order to see what they have to teach.

As for the practical domain itself, philosophers have two significant roles to play in public deliberation over environmental conflict. Given the limits of knowledge and hope, however, these roles are fairly modest: philosophers may make themselves useful, but they are not nearly as crucial to the process as they sometimes suppose.

First, philosophers can and do act as advocates for one point of view or another; as citizens, they have values and interests of their own that deserve a public hearing. Training in philosophy, in which beliefs and values are examined and re-examined, might even prepare them to offer especially sophisticated arguments on their own behalf. This does not mean that philosophers as citizens have a privileged status that allows them to dictate the terms of the debate or to decide what kind of outcome will be satisfactory. Whatever their sophistication, they can be no more sure of the wisdom of their proposals than anyone else. Furthermore, there is little reason to believe that all philosophers will (or should) come to the same conclusions about important social and political issues: among philosophers, there are as likely to be staunch defenders of private property as there are of wilderness preservation.

Second, I believe that philosophers can and should also play a broader role as participants in public deliberation as mediators or facilitators. In their training, philosophers are exposed to a wide range of worldviews and modes of argument. As a result, they may be qualified to clarify the terms of public discussion and debate, to raise important questions, to point out limits and consequences of various arguments, and perhaps to help assure that appropriate principles of skepticism—especially parity—are in play. This will not bring environmental philosophers any closer to generating the correct answer, and it does not elevate their status as ordinary participants in a much broader discussion. Instead, it confers on them the additional responsibility of helping to assure that illusion and denial are absent—as much as possible—from the process.

Notes

INTRODUCTION

1. David Hume, *An Enquiry concerning Human Understanding,* ed. Anthony Flew (LaSalle, Ill.: Open Court, 1988), 181.

2. Ibid., 192.

3. Anna Bramwell, *Ecology in the 20th Century: A History* (New Haven: Yale University Press, 1989), 23–24.

4. David Abram, *The Spell of the Sensuous: Perception and Language in a More-Than-Human World* (New York: Pantheon, 1996), 92–95.

5. Near the end of the *Critique of Pure Reason,* Immanuel Kant stated that all the interests of reason, speculative as well as practical, combine into three questions: What can I know? What ought I to do? For what may I hope? Note that the three claims of the central case answer each of these questions in turn. See Immanuel Kant, *Critique of Pure Reason,* trans. Norman Kemp Smith (New York: Macmillan, 1965), 635.

1. THE NATURE OF NATURE

1. Arne Naess, "The Shallow and the Deep, Long-Range Ecology Movements: A Summary," *Inquiry* 16 (1973): 95.

2. Bill Devall and George Sessions, *Deep Ecology: Living As If Nature Mattered* (Salt Lake City: Gibbs Smith, 1985), 42–48. See also Warwick Fox, *Toward a Transpersonal Ecology* (Albany: SUNY Press, 1995), 3–13.

3. Aldo Leopold, *A Sand County Almanac, and Sketches Here and There* (Oxford: Oxford University Press, 1949), 214–220, 224–225.

4. See, for example, Thomas Berry, *The Dream of the Earth* (San Francisco: Sierra Club Books, 1988); Murray Bookchin, *The Philosophy of Social Ecology* (New York: Black Rose Books, 1990); and Warwick Fox, *Toward a Transpersonal Ecology: Developing New Foundations for Environmentalism* (Albany: State University of New York Press, 1995), 249–258.

5. Ariel Kay Salleh, "Deeper Than Deep Ecology: The Eco-Feminist Connection," *Environmental Ethics* 6 (1984): 344; and Jim Cheney, "Eco-Feminism and Deep Ecology," *Environmental Ethics* 9 (1987): 121, 126–127.

6. Devall and Sessions, *Deep Ecology*, 65–67.

7. Fox, *Transpersonal Ecology*, 250–252.

8. See, for example, Abram, *The Spell of the Sensuous*, 31–72.

9. G. W. F. Hegel, *Philosophy of Nature*, trans. A. V. Miller (Oxford: Oxford University Press, 1970), 3.

10. Gottfried Leibniz, "Considerations on the Principle of Life, and on Plastic Natures, by the Author of the Pre-Established Harmony," in *Selections*, ed. Philip P. Wiener (New York: Charles Scribner's Sons, 1979), 196.

11. Immanuel Kant, *Critique of Judgment*, Ak. 370–374. This translation by Werner S. Pluhar (Indianapolis: Hackett Publishing Company, 1987).

12. Kant, *Critique of Judgment*, Ak. 435–436.

13. The German term *Naturphilosophie* has no equivalent in English; "philosophy of nature" is most commonly offered as a translation, but it does not capture the sense of the unity of philosophy with nature that characterized the movement. I will use "nature philosophy" as the nearest equivalent. See George Gusdorf, *Le savoir romantique de la nature* (Paris: Payot, 1985), 14–15, 21.

14. Friedrich Wilhelm Joseph von Schelling, *Ideas for a Philosophy of Nature*, trans. Errol E. Harris and Peter Heath (Cambridge: Cambridge University Press, 1988), 42.

15. Ibid., trans. A. V. Miller (Oxford: Oxford University Press, 1970), 5, 10.

16. Schelling, *Phenomenology of Spirit*, 50–51.

17. Hegel, *Ideas for a Philosophy of Nature*, 14.

18. Ibid., 444–445.

19. Ibid., 443, emphasis mine.

20. Jacques Derrida, "From Restricted to General Economy: A Hegelianism without Reserve," in *Writing and Difference*, trans. Alan Bass (Chicago: University of Chicago Press, 1978), 271.

21. Hegel, *Ideas for a Philosophy of Nature*, 292–293, 295–297.

22. Devall and Sessions, *Deep Ecology*, 80–84, 90–108.

23. Ibid., 81–82.

24. I am grateful to Steve Vogel for pointing out this distinction, which he clarifies as the difference between "nature as origin" and "nature as other."

25. See, for example, Christopher Manes, *Green Rage: Radical En-*

vironmentalism and the Unmaking of Civilization (Boston: Little, Brown and Company, 1990).

26. Regarding this example, along with a broader discussion of the social uses of nature, see Neil Evernden, *The Social Creation of Nature* (Baltimore: Johns Hopkins University Press, 1992), 3–17.

27. For detailed histories of the development of the environmentalist notion of wild nature, including Muir's arguments, see Roderick Nash, *Wilderness and the American Mind*, 3rd ed. (New Haven: Yale University Press, 1982); and Max Oelschaeger, *The Idea of Wilderness* (New Haven: Yale University Press, 1991).

2. ORGANISM AND MECHANISM

1. Devall and Sessions, *Deep Ecology*, 85; J. Baird Callicott, "The Metaphysical Implications of Ecology" *Environmental Ethics* 8 (1986): 301–302, 310–311, 313.

2. Fox, *Toward a Transpersonal Ecology*, 330, 333–335.

3. Karen J. Warren and Jim Cheney, "Ecosystem Ecology and Metaphysical Ecology: A Case Study," *Environmental Ethics* 15 (1993): 111–112.

4. On the limits of mechanism, see Jacques Roger, *Les sciences de la vie dans la pensée française au XVIIIe siècle*, 3rd ed. (Paris: Albin Michel, 1993), 206–224, 731.

5. William R. Woodward and Reinhardt Pester, "From Romantic *Naturphilosophie* to a Theory of Scientific Method for the Medical Disciplines," in *Romanticism in Science: Science in Europe, 1790–1840*, ed. Stephano Poggi and Maurizio Bossi (Dordrecht: Kluwer Academic Publishers, 1994), 164; and David Knight, "Romanticism and the Sciences," in *Romanticism and the Sciences*, ed. Andrew Cunningham and Nicholas Jardine (Cambridge: Cambridge University Press, 1990), 14.

6. Giulio Barsanti, "Lamarck and the Birth of Biology," in *Romanticism in Science*, ed. Poggi and Bossi, 55, 59.

7. François Dagognet, *Le vivant* (Paris: Bordas, 1988), 99. Translation and emphasis mine.

8. Frank N. Egerton, "Changing Concepts of the Balance of Nature," *The Quarterly Review of Biology* 48 (1973): 336.

9. Carl Linné, *L'équilibre de la nature*, ed. Camille Limoges, trans. Bernard Jasmin (Paris: J. Vrin, 1972), 58.

10. Stephen A. Forbes, "The Lake as Microcosm," reprinted in *Bulletin of the Illinois State Natural History Survey* 15 (1925): 537. A facsimile of this article can be found in Leslie A. Real and James H. Brown, eds., *Foundations of Ecology: Classic Papers with Commentaries* (Chicago: University of Chicago Press), 14–27.

11. Forbes, "The Lake as Microcosm," 537.

12. Pascal Acot characterizes ecology as a product of the *"laïcisation,"* or secularization, of natural history. See his *Histoire de l'écologie* (Paris: PUF, 1988), 14–16.

13. Linné, *L'équilibre de la nature*, 103–104.

14. Buffon, *De la manière d'étudier et de traiter l'histoire naturelle* (Paris: Société des Amis de la Bibliothèque Nationale, 1986), 10, 74. Translation mine. See also Jacques Roger, *Buffon: un philosophe au Jardin du Roi* (Paris: Fayard, 1989), 46.

15. Acot (*Histoire de l'écologie*, 46–60) marshals a wide range of evidence for Darwin's indirect influence on ecology, a feat which I will not attempt to duplicate here. James P. Collins, John Beatty, and Jane Maienschein, "Introduction: Between Ecology and Evolutionary Biology," *Journal of the History of Biology* 19 (1986): 171; and Douglas J. Futuyma, "Reflections on Reflections: Ecology and Evolutionary Biology," *Journal of the History of Biology* 19 (1986): 304–305.

16. Jean-Marc Drouin, *L'écologie et son histoire: réinventer la nature* (Paris: Flammarion, 1993), 46–47, 51.

17. Alexandre de Humboldt et Aimé Bonpland, *Essai sur la géographie des plantes* (Paris: Fr. Schoell, 1807; Nanterre: Éditions Européens Erasme, 1990), 32. See also Acot, *Histoire de l'écologie*, 18, 24, 25–42; Malcolm Nicolson, "Alexander von Humboldt and the Geography of Vegetation," in *Romanticism and the Sciences*, ed. Cunningham and Jardine, 178–181; and Roger Dajoz, "Éléments pour une histoire de l'écologie," *Histoire et Nature* 24–25 (1984): 11.

18. Nicolson, "Alexander von Humboldt," 180.

19. Collins, Beatty, and Maienschein, "Introduction: Between Ecology and Evolutionary Biology," 171.

20. Frederick Clements, *Research Methods in Ecology* (Lincoln, Nebr.: University Publishing Company, 1905), 199.

21. Ibid., 1–2, 10; Joel B. Hagen, *An Entangled Bank: The Origins of Ecosystem Ecology* (New Brunswick, N.J.: Rutgers University Press, 1992), 23.

22. H. A. Gleason, "The Individualistic Concept of the Plant Association," *Bulletin of the Torrey Botanical Club* 53 (1926): 8–9. Facsimile in *Foundations of Ecology*, ed. Real and Brown, 98–117.

23. Gleason, "The Individualistic Concept," 114; see also Hagen, *An Entangled Bank*, 28.

24. A. G. Tansley, "The Use and Abuse of Vegetational Concepts and Terms," *Ecology* 16 (1935): 285, 289–290. Facsimile in *Foundations of Ecology*, ed. Real and Brown, 318–341.

25. Tansley, "The Use and Abuse of Vegetational Concepts," 299.

26. Egerton, "Changing Concepts of the Balance of Nature," 345.

See also Clements, "The Nature and Structure of the Climax," *The Journal of Ecology* 24 (1936): 252–284. A facsimile of the Clements article is printed in *Foundations of Ecology*, ed. Real and Brown, 59–97.

27. Raymond Lindeman, "The Trophic-Dynamic Aspect of Ecology," *Ecology* 23 (1942): 157. Facsimile in *Foundations of Ecology*, ed. Real and Brown, 157–176. On the development of the metabolism metaphor under the influence of G. E. Hutchinson, see Hagen, *An Entangled Bank*, 50.

28. 'Ecosystem' was defined as "a, or the, functional unit of nature, consisting of components such as species populations (or aggregates of these) linked together by multiple interrelations, notably food webs, organic debris, inorganic minerals, atmospheric gases and of water, and flows of these, and of energy between the organic and inorganic components." Robert P. McIntosh, *The Background of Ecology: Concept and Theory* (Cambridge: Cambridge University Press, 1985), 204.

29. Eugene P. Odum, *Fundamentals of Ecology*, 3rd ed. (Philadelphia: Saunders Publishing Company, 1971), 4. Howard T. Odum tried to develop a "universal science of systems"; see Hagen, *An Entangled Bank*, 131. Note that there is a distinction to be made between "systems ecology" and "ecosystem ecology"; see McIntosh, *The Background of Ecology*, 203.

30. Hagen, *An Entangled Bank*, 138, 129; Odum, *Fundamentals of Ecology*, 32; Eugene P. Odum, "The Strategy of Ecosystem Development," *Science* 164 (1969): 262. Facsimile in *Foundations of Ecology*, ed. Real and Brown, 596–604.

31. Hagen, *An Entangled Bank*, 144.

32. R. K. Colwell, cited in McIntosh, "Pluralism in Ecology," *Annual Review of Ecology and Systematics* 18 (1987): 322.

33. McIntosh, *The Background of Ecology*, 193.

34. McIntosh, "Pluralism in Ecology," 321. See also Hagen, *An Entangled Bank*, 145.

35. This is the core of Robert Henry Peters's argument in *A Critique for Ecology* (Cambridge: Cambridge University Press, 1991).

36. See McIntosh, "Pluralism in Ecology," 321–341. For the variety of definitions given to key terms in ecology, see K. S. Shrader-Frechette and E. D. McCoy, *Method in Ecology: Strategies for Conservation* (Cambridge: Cambridge University Press, 1993), 61–67. The two tables on these pages are part of a broader critical overview of the epistemological status of ecological concepts and theories.

37. Daniel Botkin, *Discordant Harmonies* (New York: Oxford University Press, 1990), 124.

38. Hume, *An Enquiry concerning Human Understanding*, 195.

39. Frank B. Golley, "Deep Ecology from the Perspective of Ecological Science," *Environmental Ethics* 9 (1987): 52; Don R. Marietta, Jr., "The Interrelationship of Ecological Science and Environmental Ethics" *Environmental Ethics* 1 (1979): 201, 205, 207.

40. J. Baird Callicott, "Rolston on Intrinsic Value: A Deconstruction," *Environmental Ethics* 14 (1992): 139, 142.

41. Devall and Sessions, *Deep Ecology,* 66, 80, 85–90.

42. Naess, "The Shallow and the Deep," 98–99 (emphasis removed); Golley, "Deep Ecology," 46–47, 52.

43. Abram, *The Spell of the Sensuous,* 65. See also Abram, "Merleau-Ponty and the Voice of the Earth," *Environmental Ethics* 10 (1988): 106.

44. Abram, *The Spell of the Sensuous,* 84–85.

45. Thomas Berry, *The Dream of the Earth,* 16.

46. Stephen Jay Gould, *Full House: The Spread of Excellence from Plato to Darwin* (New York: Harmony Books, 1996), 17–29. Gould refers to Freud's reflection that the advancement of knowledge since Copernicus has produced a series of attacks on human narcissism and self-assurance. Gould explores the Copernican, Darwinian, and Freudian revolutions; he also explores the consequences of the discovery of "deep time" by geologists and paleontologists.

3. A PLACE ON EARTH

1. Jean-Jacques Rousseau, *Oevres complètes de Jean-Jacques Rousseau,* edited for the Bibliothèque de la Pléiade by B. Gagnebin, M. Raymond, et al. 5 volumes. (Paris: Éditions Gallimard, 1959–1995), I: 1044. (Hereafter *OC;* all translations my own).

2. Leopold, *A Sand County Almanac,* 202–204, 224–225.

3. Rousseau, *OC,* III: 365.

4. Rousseau, *OC,* III: 207.

5. Rousseau, *OC,* IV: 856.

6. Lawrence E. Joseph, *Gaia: The Growth of an Idea* (New York: St. Martin's Press, 1990), 67–68.

7. Neil Evernden, *The Natural Alien: Humankind and Environment* (Toronto: University of Toronto Press, 1985), 39, 42–43, 74–75.

8. Rousseau, *OC,* III: 139.

9. Rousseau, *OC,* III: 414–419.

10. Rousseau, *OC,* I: 1062.

11. Rousseau, *OC,* I: 566, 1083.

12. Stephen A. Forbes, "The Humanizing of Ecology," *Ecology* 3 (1922): 89–90.

13. McIntosh, *The Background of Ecology,* 307.

14. See Paul Shepard and Daniel McKinley, eds., *The Subversive*

Science: Essays Toward an Ecology of Man (Boston: Houghton Mifflin, 1969), vii.

15. Clarence J. Glacken, *Traces on the Rhodian Shore: Nature and Culture in Western Thought from Ancient Times to the End of the Eighteenth Century* (Berkeley: University of California Press, 1967), vii, 80–81. Glacken goes on to discuss the Hippocratic text *Airs, Waters, Places* in more detail, raising questions of authorship and pointing out some difficulties with the theory. He notes that Hippocrates is more interested in explaining the differences between cultures than the similarities among them (85). He also notes that the text, through its widespread influence, "is responsible for the fallacy that, if environmental influences on the physical and mental qualities of individuals can be shown, they can by extension be applied to whole peoples" (88).

16. Later published as *Man's Place in Nature* (Ann Arbor: Ann Arbor Paperbacks, 1959).

17. Charles Darwin, *The Descent of Man and Selection in Relation to Sex* (Princeton: Princeton University Press, 1981); and Darwin, *Expression of the Emotions in Man and Animals* (New York: Philosophical Library, 1955).

18. Darwin, *The Descent of Man,* 53–65, 71–72.

19. McIntosh, *The Background of Ecology,* 307.

20. Robert E. Park, "The City: Suggestions for the Investigation of Human Behavior in the Urban Environment," in Robert E. Park, Ernest W. Burgess, and Roderick D. McKenzie, *The City* (Chicago: University of Chicago Press, 1925), 1.

21. Roderick D. McKenzie, "The Ecological Approach to the Study of Human Community," in Park, Burgess, and McKenzie, *The City,* 63–65, 68.

22. Shepard and McKinley, *The Subversive Science,* vii. For a sampling of the range of work included in the volume, see also V. C. Wynne-Edwards, "Self-Regulating Systems in Populations of Animals," 99–111; Frank E. Egler, "Pesticides—in Our Ecosystem," 245–267, especially the illustrations of "knowledge flow" and "information flow" on 260, 262, 264; Garrett Hardin, "The Cybernetics of Competition: a Biologist's View of Society," 275–296; and F. Fraser Darling, "The Ecological Approach to the Social Sciences," 316–327.

23. I have developed this analogy more thoroughly in my paper "Why Ecology Cannot Be All Things to All People: The 'Adaptive Radiation' of Scientific Concepts," *Environmental Ethics* 19 (1997): 375–390.

24. See Acot, *Histoire de l'écologie,* 174.

25. Leopold, *A Sand County Almanac,* 202.

26. Callicott, "Hume's *Is/Ought* Dichotomy and the Relation of Ecology to Leopold's Land Ethic" *Environmental Ethics* 4 (1982): 165–166, 173–174.

27. Kirkpatrick Sale, *Dwellers in the Land: The Bioregional Vision* (San Francisco: Sierra Club Books, 1995), 54–55.

28. Edward S. Casey. *Getting Back into Place: Toward a Renewed Understanding of the Place-World* (Bloomington: Indiana University Press, 1993), xiii–xv.

29. Sale, *Dwellers in the Land*, 55.

30. Casey, *Getting Back into Place*, 31.

31. Deane Curtin, *Chinnagounder's Challenge: The Question of Ecological Citizenship* (Bloomington: Indiana University Press, 1999), 158–159.

4. THE MORAL COMPASS

1. Leopold, *A Sand County Almanac*, 202–204, 224–225.

2. Robert Elliot, *Faking Nature: The Ethics of Environmental Restoration* (New York: Routledge, 1997), 5–11, 34.

3. J. Baird Callicott, "The Case against Moral Pluralism," *Environmental Ethics* 12 (1990): 99–124.

4. Bryan G. Norton, "Why I Am Not a Nonanthropocentrist: Callicott and the Failure of Monistic Inherentism," *Environmental Ethics* 17 (1995): 345. See also Anthony Weston, "On Callicott's Case against Moral Pluralism," *Environmental Ethics* 13 (1991): 283–286.

5. Paul Taylor, *Respect for Nature* (Princeton: Princeton University Press, 1986), 80.

6. Holmes Rolston, III, *Environmental Ethics: Duties to and Values in the Natural World* (Philadelphia: Temple University Press, 1988), 98–101.

7. See Peter Singer, *Animal Liberation*, 2nd ed. (New York: New York Review of Books, 1990); and Tom Regan, *The Case for Animal Rights* (Berkeley: University of California Press, 1983).

8. Leopold, *A Sand County Almanac*, 224. Emphasis mine.

9. Fox, *Transpersonal Ecology*, 217; Naess, "The Shallow and the Deep," 95–96.

10. Karen J. Warren, "The Power and the Promise of Ecological Feminism," in *Ecological Feminist Philosophies*, ed. Karen J. Warren (Bloomington: Indiana University Press, 1996), 32–33.

11. Callicott, "Hume's *Is/Ought* Dichotomy," 165–166, 173–174.

12. Rolston, *Environmental Ethics*, 230–232.

13. Marietta, "The Interrelationship of Ecological Science and Environmental Ethics," 200–203.

14. Devall and Sessions, *Deep Ecology*, 99.

15. Bruce V. Foltz, *Inhabiting the Earth: Heidegger, Environmental Ethics, and the Metaphysics of Nature* (Atlantic Highlands, N.J.: Humanities Press, 1995), 172.

16. Ibid., 173.

17. Ibid., 22.

18. Donald Worster, *Nature's Economy: A History of Ecological Ideas* (New York: Cambridge University Press, 1977), 9.

19. George Perkins Marsh, *Man and Nature or, Physical Geography As Modified by Human Action* (Cambridge: Belknap Press, 1965), 43, 36.

20. Worster, *Nature's Economy*, 39–50.

21. See Roger, *Buffon*, 312–313.

22. Acot, *Histoire de l'écologie*, 84.

23. Hagen, *An Entangled Bank*, 101, 107–112, 115–118.

24. Ibid., 101–107, 112–115.

25. Foltz, *Inhabiting the Earth*, 166.

26. J. Baird Callicott, "The Search for an Environmental Ethic," in *Matters of Life and Death: New Introductory Essays in Moral Philosophy*, ed. Tom Regan. 3rd ed. (New York: McGraw-Hill, 1993), 333, 335–337.

27. Elliot, *Faking Nature*, 30, 117.

5. Environmentalism without Illusions

1. Wendell Berry, "Why I Am Not Going to Buy a Computer," in *What Are People For?* (San Francisco: North Point Press, 1990), 177; and Anthony Weston, *Back to Earth: Tomorrow's Environmentalism* (Philadelphia: Temple University Press, 1994), 11–12.

2. George Lakoff and Mark Johnson, *Metaphors We Live By* (Chicago: University of Chicago Press, 1980), 143.

3. Ibid., 144.

4. This is the logical culmination of the trend Aldo Leopold spotted in the 1940s. See *A Sand County Almanac*, 165–166.

5. Ibid., 205, 214. The interpretation of Leopold is still contested territory among environmental philosophers. See Bryan G. Norton, *Toward Unity among Environmentalists* (Oxford: Oxford University Press, 1991), 57–60; and compare it with J. Baird Callicott, "The Conceptual Foundations of the Land Ethic," in *In Defense of the Land Ethic* (Albany: SUNY Press, 1989), 75–99.

6. See Devall and Sessions, *Deep Ecology*, 227.

7. Jim Cheney, "Postmodern Environmental Ethics: Ethics as Bioregional Narrative," *Environmental Ethics* 11 (1989): 117–134.

8. Andrew Light and Eric Katz, "Introduction: Environmental Pragmatism and Environmental Ethics as Contested Terrain," in *Environmental Pragmatism*, ed. Andrew Light and Eric Katz (New York:

Routledge, 1996), 15–16. Norton, *Toward Unity among Environmentalists,* 11–13.

9. Eugene Hargrove, "After Fifteen Years," *Environmental Ethics* 15 (1993): 292; Hargrove, "After Twenty Years," *Environmental Ethics* 20 (1998): 340.

10. See Botkin, *Discordant Harmonies,* 16–19.

11. In the language of negotiation, each participant has a BATNA, or "Best Alternative To a Negotiated Agreement," which is the standard against which any proposed settlement is measured. See Roger Fisher, William Ury, and Bruce Patton, *Getting to Yes: Negotiating Agreement without Giving In,* 2nd ed. (New York: Penguin, 1981), 100.

12. Weston, *Back to Earth,* 12, 14. Robert Mugerauer once told me that environmental philosophers should "let a thousand flowers bloom." I think he meant something like this.

13. David Schlosberg, *Environmental Justice and the New Pluralism: The Challenge of Difference for Environmentalism* (Oxford: Oxford University Press, 1999), 3–4.

14. Curtin, *Chinnagounder's Challenge,* 155, 164.

15. Robert D. Bullard, "Environmental Justice for All," in *Unequal Protection: Environmental Justice and Communities of Color,* ed. Robert D. Bullard (San Francisco: Sierra Club Books, 1994), 7–12; and Schlosberg, *Environmental Justice and the New Pluralism,* 107–108. Bullard provides an excellent introduction to environmental justice, as does David E. Newton in *Environmental Justice* (Santa Barbara: ABC–CLIO, 1996). Schlosberg argues that the environmental justice movement embodies a renewed commitment to pluralism.

16. The conflict/dispute distinction is attributed to John Burton. Guy Burgess and Heidi Burgess, "Beyond the Limits: Dispute Resolution of Intractable Environmental Conflicts," in *Mediating Environmental Conflicts: Theory and Practice,* ed. J. Walton Blackburn and Willa Marie Bruce (Westport, Conn.: Quorum Books, 1995), 101–119. Emphasis removed.

17. Curtin, *Chinnagounder's Challenge,* 164. Emphasis removed.

18. Aldo Leopold, *A Sand County Almanac,* 68.

19. Bryan G. Norton, "Applied Philosophy versus Practical Philosophy: Toward an Environmental Policy Integrated According to Scale," in *Environmental Philosophy and Environmental Activism,* ed. Don E. Marietta, Jr., and Lester Embree (Lanham, Md.: Rowman and Littlefield, 1995), 126.

20. Norton, *Toward Unity among Environmentalists,* 187–204.

Bibliography

Abram, David. "Merleau-Ponty and the Voice of the Earth." *Environmental Ethics* 10 (1988): 101–120.

———. *The Spell of the Sensuous: Language and Meaning in a More-Than-Human World.* New York: Pantheon, 1996.

Acot, Pascal. "Écologie et écologisme." *Raison Présente*, no. 106 (1993): 27–36.

———. *Histoire de l'écologie.* Paris: PUF, 1988.

Adams, Charles C. "The Relation of General Ecology to Human Ecology." *Ecology* 16 (1935): 316–335.

Aristotle. *Generation of Animals.* Translated by A. L. Peck. Cambridge, Mass.: Harvard University Press, 1942.

———. *Parts of Animals.* Rev. ed. Translated by A. L. Peck and E. S. Forster. Cambridge, Mass.: Harvard University Press, 1961.

Ayer, Alfred Jules. *Language, Truth and Logic.* New York: Dover Publications, 1952.

Bachelard, Gaston. *Épistémologie: textes choisis.* Edited by Dominique Lecourt. Paris: PUF, 1971.

———. *La formation de l'esprit scientifique.* Paris: J. Vrin, 1993.

Balan, Bernard. "Premières recherches sur l'origine et la formation du concept d'économie animale." *Revue d'Histoire des Sciences* 28 (1975): 289–326.

Barsanti, Giulio. "Linné et Buffon: deux visions différentes de la nature et de l'histoire naturelle." *Révue de Synthèse*, 3rd ser., 105 (1984): 83–112.

Bates, Marston. *The Nature of Natural History.* Princeton: Princeton University Press, 1990.

Bergandi, Donato. "*Fundamentals of ecology* de E.P. Odum: veritable 'approche holiste' ou reductionisme masqué?" *Bulletin d'Écologie* 24 (1992): 57–68.

Berry, Thomas. *The Dream of the Earth*. San Francisco: Sierra Club Books, 1988.

Blackburn, J. Walton, and Willa Marie Bruce, eds. *Mediating Environmental Conflicts: Theory and Practice*. Westport, Conn.: Quorum Books, 1995.

Bookchin, Murray. *The Philosophy of Social Ecology: Essays on Dialectical Naturalism*. Montreal: Black Rose Books, 1990.

Botkin, Daniel B. *Discordant Harmonies: A New Ecology for the Twenty-First Century*. Oxford: Oxford University Press, 1990.

Bramwell, Anna. *Ecology in the Twentieth Century: A History*. New Haven: Yale University Press, 1989.

Buffon, George Louis Leclerc, comte de. *De la manière d'étudier et de traiter l'histoire naturelle*. Paris: Société des Amis de la Bibliothèque Nationale, 1986.

———. *De l'homme*. Edited by Michèle Duchet. Paris: François Maspero, 1971.

———. *L'histoire naturelle de l'homme et des animaux*. Paris: J. de Bonnot, 1989.

———. *Oeuvres philosophiques de Buffon*. Edited by Jean Piveteau. Paris: PUF, 1954.

Bullard, Robert D., ed. *Unequal Protection: Environmental Justice and Communities of Color*. San Francisco: Sierra Club Books, 1994.

Callicott, J. Baird. "The Case against Moral Pluralism." *Environmental Ethics* 12 (1990): 99–124.

———. "Hume's *Is/Ought* Dichotomy and the Relation of Ecology to Leopold's Land Ethic." *Environmental Ethics* 4 (1982): 163–174.

———. "The Metaphysical Implications of Ecology." *Environmental Ethics* 8 (1986): 301–316.

———. "Rolston on Intrinsic Value: A Deconstruction." *Environmental Ethics* 14 (1992): 129–143.

Canguilhem, Georges. *La connaissance de la vie*. Paris: J. Vrin, 1965.

———. *Ideology and Rationality in the History of the Life Sciences*. Translated by Arthur Goldhammer. Cambridge, Mass.: MIT Press, 1988.

Casey, Edward S. *Getting Back into Place: Toward a Renewed Understanding of the Place-World*. Bloomington: Indiana University Press, 1993.

Cheney, Jim. "Callicott's 'Metaphysics of Morals.'" *Environmental Ethics* 13 (1991): 311–325.

———. "Eco-Feminism and Deep Ecology." *Environmental Ethics* 9 (1987): 115–145.

———. "Postmodern Environmental Ethics: Ethics as Bioregional Narrative." *Environmental Ethics* 11 (1989): 117–134.

Clements, Frederic Edward. *Research Methods in Ecology*. Lincoln, Nebr.: University Publishing Company, 1905.

Collingwood, R. G. *The Idea of Nature*. Oxford: Oxford University Press, 1960.

Collins, James P., John Beatty, and Jane Maienschein. "Introduction: Between Ecology and Evolutionary Biology." *Journal of the History of Biology* 19 (1986): 169–180.

Cunningham, Andrew, and Nicholas Jardine, eds. *Romanticism and the Sciences*. Cambridge: Cambridge University Press, 1990.

Curtin, Deane. *Chinnagounder's Challenge: The Question of Ecological Citizenship*. Bloomington: Indiana University Press, 1999.

Dagognet, François. *Nature*. Paris: J. Vrin, 1990.

——. *Le vivant*. Paris: Bordas, 1988.

Dajoz, Roger. "Éléments pour une histoire de l'écologie." *Histoire et Nature*, nos. 24–25 (1984): 5–112.

Darwin, Charles. *The Descent of Man and Selection in Relation to Sex*. Princeton: Princeton University Press, 1981.

——. *Expression of the Emotions in Man and Animals*. New York: Philosophical Library, 1955.

——. *The Origin of Species*. London: John Murray, 1859; Cambridge, Mass.: Harvard University Press, 1964.

Davis, Donald Edward. *Ecophilosophy: A Field Guide to the Literature*. San Pedro, Calif.: R. & E. Miles, 1989.

Deléage, Jean-Paul. *Histoire de l'écologie: une science de l'homme et de la nature*. Paris: La Découverte, 1991.

Derrida, Jacques. *Writing and Difference*. Translated by Alan Bass. Chicago: University of Chicago Press, 1978.

Descartes, René. *Discours de la méthode*. Paris: Garnier-Flammarion, 1996.

Devall, Bill, and George Sessions. *Deep Ecology: Living As If Nature Mattered*. Salt Lake City: Gibbs Smith, 1985.

Di Castri, Francesco. *L'écologie: les défis d'une science en temps de crise*. Paris: La Documentation Française, 1984.

Drouin, Jean-Marc. *L'écologie et son histoire: réenventer la nature*. Paris: Flammarion, 1993.

Egerton, Frank N. "A Bibliographical Guide to the History of General Ecology and Population Ecology." *History of Science* 15 (1977): 189–215.

——. "Changing Concepts of the Balance of Nature." *The Quarterly Review of Biology* 48 (1973): 322–350.

——. "The History of Ecology: Achievements and Opportunities." *Journal of the History of Biology* 16 (1983): 259–310; 18 (1985): 103–143.

Elliot, Robert. *Faking Nature: The Ethics of Environmental Restoration*. New York: Routledge, 1997.

Evernden, Neil. *The Natural Alien: Humankind and Environment.* Toronto: University of Toronto Press, 1985.

——. *The Social Creation of Nature.* Baltimore: Johns Hopkins University Press, 1992.

Ferré, Frederick, and Peter Hartel, eds. *Ethics and Environmental Policy: Theory Meets Practice.* Athens: University of Georgia Press, 1994.

Ferry, Luc. *The New Ecological Order.* Translated by Carol Volk. Chicago: University of Chicago Press, 1995. Originally published as *Le nouvel ordre écologique: l'arbre, l'animal, l'homme.* Paris: Grasset, 1992.

Feyerabend, Paul. *Against Method.* 3rd ed. New York: Verso, 1993.

Foltz, Bruce V. *Inhabiting the Earth: Heidegger, Environmental Ethics, and the Metaphysics of Nature.* Atlantic Highlands, N.J.: Humanities Press, 1995.

Forbes, Stephen A. "The Humanizing of Ecology." *Ecology* 3 (1922): 89–92.

Fox, Warwick. "The Deep Ecology–Ecofeminism Debate and Its Parallels." *Environmental Ethics* 11 (1989): 5–25.

——. *Toward a Transpersonal Ecology: Developing New Foundations for Environmentalism.* Albany: State University of New York Press, 1995.

Futuyma, Douglas J. "Reflections on Reflections: Ecology and Evolutionary Biology." *Journal of the History of Biology* 19 (1986): 302–312.

Glacken, Clarence. *Traces on the Rhodian Shore: Nature and Culture in Western Thought from Ancient Times to the End of the Eighteenth Century.* Berkeley: University of California Press, 1967.

Goldsmith, Edward. *The Way: An Ecological World-view.* Boston: Shambhala, 1993.

Golley, Frank B. "Deep Ecology from the Perspective of Ecological Science." *Environmental Ethics* 9 (1987): 45–55.

Gould, Stephen Jay. *Full House: The Spread of Excellence from Plato to Darwin.* New York: Harmony Books, 1996.

Gusdorf, George. *Le savoir romantique de la nature.* Paris: Payot, 1985.

Hagen, Joel B. *An Entangled Bank: The Origins of Ecosystem Ecology.* New Brunswick, N.J.: Rutgers University Press, 1992.

Hargrove, Eugene C. "After Fifteen Years." *Environmental Ethics* 15 (1993): 292.

——. "After Twenty Years." *Environmental Ethics* 20 (1998): 340.

——. *Foundations of Environmental Ethics.* Denton, Tex.: Environmental Ethics Books, 1989.

Hegel, Georg Wilhelm Friedrich. *The Encyclopedia Logic.* Translated by T. F. Geraets, W. A. Suchting, and H. S. Haris. Indianapolis: Hackett Publishing Company, 1991.

————. *The Encyclopedia of the Philosophical Sciences in Outline*. Edited by Ernst Behler. New York: Continuum, 1990.

————. *Phenomenology of Spirit*. Translated by A. V. Miller. Oxford: Oxford University Press, 1970.

————. *The Philosophy of Nature*. Translated by A. V. Miller. Oxford: Oxford University Press, 1970.

Hempel, Carl G. *Philosophy of Natural Science*. Englewood Cliffs, N.J.: Prentice Hall, 1966.

Humboldt, Alexandre, and Aimé Bonpland. *Essai sur la géographie des plantes*. Paris: Fr. Schoell, 1807; Nanterre: Éditions Européens Erasme, 1990.

Hume, David. *An Enquiry concerning Human Understanding*. Edited by Anthony Flew. LaSalle, Ill.: Open Court, 1988.

Husserl, Edmund. *The Crisis of European Sciences and Transcendental Phenomenology*. Translated by David Carr. Evanston: Northwestern University Press, 1970.

Hutchinson, George Evelyn. *The Ecological Theatre and the Evolutionary Play*. New Haven: Yale University Press, 1965.

Huxley, Thomas Henry. *Man's Place in Nature*. Ann Arbor: Ann Arbor Paperbacks, 1959.

Joseph, Lawrence E. *Gaia: The Growth of an Idea*. New York: St. Martin's Press, 1990.

Kant, Immanuel. *Critique of Judgment*. Translated by Werner S. Pluhar. Indianapolis: Hackett Publishing Company, 1987.

————. *Critique of Pure Reason*. Translated by Norman Kemp Smith. New York: Macmillan, 1965.

Kirkman, Robert. "Why Ecology Cannot Be All Things to All People: The 'Adaptive Radiation' of Scientific Concepts." *Environmental Ethics* 19 (1997): 375–390.

Kuhn, Thomas S. *The Structure of Scientific Revolutions*. 2nd ed. Chicago: University of Chicago Press, 1970.

Lakatos, Imre, and Alan Musgrave, eds. *Criticism and the Growth of Knowledge*. Cambridge: Cambridge University Press, 1970.

Lakoff, George, and Mark Johnson. *Metaphors We Live By*. Chicago: University of Chicago Press, 1980.

Leibniz, Gottfried. *Selections*. Edited by Philip P. Wiener. New York: Charles Scribner's Sons, 1979.

Leopold, Aldo. *A Sand County Almanac and Sketches Here and There*. Oxford: Oxford University Press, 1949.

Levins, Richard, and Richard Lewontin. "Dialectics and Reductionism in Ecology." *Synthèse* 43 (1980): 47–78.

Light, Andrew, and Eric Katz, eds. *Environmental Pragmatism*. New York: Routledge, 1996.

Lindley, David. "Is the Earth Alive or Dead?" *Nature* 332 (1988): 483–484.

Linné, Carl. *L'équilibre de la nature.* Edited by Camille Limoges. Translated by Bernard Jasmin. Paris: J. Vrin, 1972.

List, Peter C. *Radical Environmentalism: Philosophy and Tactics.* Belmont, Calif.: Wadsworth, 1993.

Lovelock, James E. *The Ages of Gaia: A Biography of Our Living Earth.* New York: W. W. Norton, 1988.

———. *Gaia: A New Look at Life on Earth.* Oxford: Oxford University Press, 1987.

Manes, Christopher. *Green Rage: Radical Environmentalism and the Unmaking of Civilization.* Boston: Little, Brown and Company, 1990.

Marietta, Jr., Don E. "The Interrelationship of Ecological Science and Environmental Ethics." *Environmental Ethics* 1 (1979): 195–207.

———. "Knowledge and Obligation in Environmental Ethics: A Phenomenological Analysis." *Environmental Ethics* 4 (1982): 153–162.

———. "World Views and Moral Decisions: A Reply to Tom Regan." *Environmental Ethics* 2 (1980): 369–371.

Marietta, Jr., Don E., and Lester Embree, eds. *Environmental Philosophy and Environmental Activism.* Lanham, Md.: Rowman and Littlefield, 1995.

Marsh, George Perkins. *Man and Nature; or, Physical Geography As Modified by Human Action.* 1864; Cambridge, Mass.: Belknap Press of Harvard University Press, 1965.

May, Robert M. "The Role of Theory in Ecology." *American Zoologist* 21 (1981): 903–910.

May, Robert M., and John Seger. "Ideas in Ecology." *American Scientist* 74 (1986): 256–267.

Mayr, Ernst. *The Growth of Biological Thought.* Cambridge, Mass.: Belknap Press, 1982.

McIntosh, Robert P. *The Background of Ecology: Concept and Theory.* Cambridge: Cambridge University Press, 1985.

———. "Pluralism in Ecology." *Annual Review of Ecology and Systematics* 18 (1987): 321–341.

Merchant, Carolyn. *The Death of Nature: Women, Ecology, and the Scientific Revolution.* New York: Harper and Row, 1980.

———. *Radical Ecology: The Search for a Livable World.* New York: Routledge, 1992.

Naess, Arne. "A Defense of the Deep Ecology Movement." *Environmental Ethics* 6 (1984): 265–70.

———. *Ecology, Community, and Lifestyle: Outline of an Ecosophy.* Translated and edited by David Rothenberg. New York: Cambridge University Press, 1989.

———. "The Shallow and the Deep, Long-Range Ecology Movements: A Summary." *Enquiry* 16 (1973): 95–100.

Nash, Roderick. *Wilderness and the American Mind*. 3rd ed. New Haven: Yale University Press, 1982.

Newton, David E. *Environmental Justice*. Santa Barbara, Calif.: ABC–CLIO, 1996.

Nietzsche, Friedrich. *The Gay Science*. Translated by Walter Kaufmann. New York: Vintage, 1974.

Norton, Bryan G. *Toward Unity among Environmentalists*. Oxford: Oxford University Press, 1991.

———. "Why I Am Not a Nonanthropocentrist: Callicott and the Failure of Monistic Inherentism." *Environmental Ethics* 17 (1995): 341–358.

Odum, Eugene P. "The Emergence of Ecology as a New Integrative Discipline." *Science* 25 (1977): 1289–1293.

———. *Fundamentals of Ecology*. 3rd ed. Philadelphia: Saunders College Publishing, 1971.

Oelschlaeger, Max. *The Idea of Wilderness: From Prehistory to the Age of Ecology*. New Haven: Yale University Press, 1991.

Park, Robert E., Ernest W. Burgess, and Roderick D. McKenzie. *The City*. Chicago: University of Chicago Press, 1967.

Peters, Robert Henry. *A Critique for Ecology*. Cambridge: Cambridge University Press, 1991.

———. "From Natural History to Ecology." *Perspectives in Biology and Medicine* 24 (1980): 191–203.

Poggi, Stefano, and Maurizio Bossi, eds. *Romanticism in Science: Science in Europe, 1890–1840*. Dordrecht: Kluwer Academic Publishers, 1994.

Popper, Karl R. *Conjectures and Refutations: The Growth of Scientific Knowledge*. 5th ed. New York: Routledge, 1989.

———. *The Logic of Scientific Discovery*. New York: Routledge, 1992.

Real, Leslie A., and James H. Brown, eds. *Foundations of Ecology: Classic Papers with Commentaries*. Chicago: University of Chicago Press, 1991.

Regan, Tom. *The Case for Animal Rights*. Berkeley: University of California Press, 1983.

———. "On the Connection between Environmental Science and Environmental Ethics." *Environmental Ethics* 2 (1980): 363–367.

Roger, Jacques. *Buffon: un philosophe au Jardin du Roi*. Paris: Fayard, 1989.

———. *Les sciences de la vie dans la pensée française au XVIIIe siècle*. 3rd ed. Paris: Albin Michel, 1993.

Rolston, III, Holmes. *Environmental Ethics: Duties to and Values in the Natural World*. Philadelphia: Temple University Press, 1988.

Rousseau, Jean-Jacques. *Oeuvres complètes de Jean-Jacques Rousseau*. Edited for the Bibliothèque de la Pléiade by B. Gagnebin, M. Raymond, et al. 5 volumes. Paris: Éditions Gallimard, 1959–1995.

Ruse, Michael. *Philosophy of Biology Today*. Albany: State University of New York Press, 1988.

Sale, Kirkpatrick. *Dwellers in the Land: The Bioregional Vision*. San Francisco: Sierra Club Books, 1995.

Salleh, Ariel Kay. "Deeper Than Deep Ecology: The Eco-Feminist Connection." *Environmental Ethics* 6 (1984): 339–345.

Schlosberg, David. *Environmental Justice and the New Pluralism: The Challenge of Difference for Environmentalism*. Oxford: Oxford University Press, 1999.

Schelling, Friedrich Wilhelm Joseph von. *Ideas for a Philosophy of Nature*. Translated by Errol E. Harris and Peter Heath. Cambridge: Cambridge University Press, 1988.

Schneider, Stephen H. "Debating Gaia." *Environment* 32, no. 4 (May 1990): 5–9, 29–32.

Schneider, Stephen H., and Penelope J. Boston, eds. *Scientists on Gaia*. Cambridge, Mass.: MIT Press, 1991.

Seamon, David, and Robert Mugerauer, eds. *Dwelling, Place, and Environment: Toward a Phenomenology of Person and World*. New York: Columbia University Press, 1989.

Shepard, Paul. "Whatever Happened to Human Ecology?" *BioScience* 17 (1967): 891–894.

Shepard, Paul, and Daniel McKinley, eds. *The Subversive Science: Essays toward an Ecology of Man*. Boston: Houghton Mifflin, 1969.

Shrader-Frechette, K. S., and E. D. McCoy. *Method in Ecology: Strategies for Conservation*. Cambridge: Cambridge University Press, 1993.

Simberloff, David. "A Succession of Paradigms in Ecology: Essentialism to Materialism and Probabilism." *Synthèse* 43 (1980): 3–39.

Singer, Peter. *Animal Liberation*. 2nd ed. New York: New York Review of Books, 1990.

Slicer, Deborah. "Is There an Ecofeminism–Deep Ecology 'Debate'?" *Environmental Ethics* 17 (1995): 151–169.

Slobodkin, Lawrence B. "Intellectual Problems of Applied Ecology." *BioScience* 38 (1988): 337–342.

Stauffer, Robert Clinton. "Ecology in the Long Manuscript Version of Darwin's *Origin of Species* and Linnaeus' *Oeconomy of Nature*." *Proceedings of the American Philosophical Society* 104 (1960): 235–241.

Steverson, Bryan K. "Ecocentrism and Ecological Modeling." *Environmental Ethics* 16 (1994): 71–88.

Swimme, Brian. *The Universe Is a Green Dragon: A Cosmic Creation Story*. Santa Fe: Bear and Company, 1984.

Taton, René, ed. *Histoire Générale des Sciences.* Paris: PUF, 1969.

Taylor, Paul. *Respect for Nature.* Princeton: Princeton University Press, 1986.

Warren, Karen J., ed. *Ecological Feminist Philosophies.* Bloomington: Indiana University Press, 1996.

Warren, Karen J., and Jim Cheney. "Ecosystem Ecology and Metaphysical Ecology: A Case Study." *Environmental Ethics* 15 (1993): 99–116.

Weston, Anthony. *Back to Earth: Tomorrow's Environmentalism.* Philadelphia: Temple University Press, 1994.

———. "On Callicott's Case against Moral Pluralism." *Environmental Ethics* 13 (1991): 283–286.

White, Jr., Lynne. "The Historical Roots of Our Ecologic Crisis." *Science* 155 (1967): 1203–1207.

Williams, George C. "*Gaia,* Nature Worship, and Biocentric Fallacies." *The Quarterly Review of Biology* 67 (1992): 479–486.

Wittbecker, Alan. "Metaphysical Implications from Physics and Ecology." *Environmental Ethics* 12 (1990): 275–282.

Worster, Donald. *Nature's Economy: A History of Ecological Ideas.* Cambridge: Cambridge University Press, 1977.

Zimmerman, Michael E. "Feminism, Deep Ecology and Environmental Ethics." *Environmental Ethics* 9 (1987): 21–44.

Index

Abram, David, 7; ambivalence of, toward the natural sciences, 79–81
adequacy, 142, 151; as litmus test, 126, 144; problems with, 162–163, 166; strong versus weak, 143–144, 145–146
animal rights, 123–124
anthropocentrism: and adequacy, 144; as foil for environmental ethics, 119–120; and purposive evolution, 82; in skeptical environmentalism, 171–173; and speculation, 38, 99 (*see also* anthropomorphism); varieties of, 172–173
anthropomorphism, 38, 97, 99
arcadian tradition, 132; doubts about, 138
Atomic Energy Commission, 136
authenticity: in environmental thought, 141–142; and place, 116; risks of, 168

balance of nature: doubts about, 70–71, 137–138; in ecology, 63; intrinsic value of, 123, 132; in Linnaean economy, 56, 59
Berry, Thomas, 83. *See also* universe story
Berry, Wendell, 150
biocentrism, 122; attributed to Heidegger, 128; and ecocentrism, 123; and human interests, 124
biology, 54–55; organicism and mechanism in, 55–56

bioregionalism, 114
biosphere, 80–81
biotic community: in bioregionalism, 114; in Clements's work, 64–65; in Forbes's work, 57; human community as, 108; in human ecology, 110; in Leopold's land ethic, 90–91, 110–111, 118–119. *See also* ecology; ecosystem
Botkin, Daniel, 70–71
Bramwell, Anna, 6–7
Buffon, George Louis Leclerc, Comte de: ethical stance of, 136, 139; on human nature, 106–107; and imperialist tradition, 133; and secularization of natural history, 59–60
Bullard, Robert, 175

Callicott, J. Baird: on deep ecology, 51; on ecofeminist ethics, 143; on ecology, 49–50; on is/ought dichotomy, 127; on moral sentiments, 111–113; and natural sciences, 77–78
Casey, Edward, 115, 116–117
causality: in Hume's skepticism, 2–3; as reconceived by Kant, 32, 33–35. *See also* final causality; mechanism
chemical metaphor, 152; limits of, 155
Cheney, Jim: postmodern environmental ethics of, 163–164; versus Callicott, 51–52, 143
Chicago school of urban sociology, 108–109

ROBERT KIRKMAN is Assistant Professor of Science and Technology Studies at the Lyman Briggs School at Michigan State University. His current research interests encompass environmental philosophy, the history and philosophy of science, the history of philosophy, and suburban environments.